THE
NUMBER
GAMES

Adam Spencer

THE
NUMBER
GAMES

XOUM PUBLISHING

Sydney

XOUM

First published by Xoum in 2017

Xoum Publishing
PO Box Q324, QVB Post Office,
NSW 1230, Australia
www.xoum.com.au

ISBN 978-1-925143-88-1 (print)
ISBN 978-1-925589-18-4 (eBook)

Cataloguing-in-Publication data is available from the National Library of Australia

Cover design, text design and typesetting by Xou Creative, www.xou.com.au

Author photograph by David Stefanoff
Printed and bound through Asia Pacific Offset

To Ellie, Olivia and Yana, three women
I love even more than mathematics!

And to William Hay, who urged our
young minds to know no limits.

Welcome!

It's that time of year again, when, after months of compiling all sorts of trivia and mathematics and hundreds of hours of writing it all up, I stare at a blank computer screen attempting to find the words that capture exactly what this book is about!

The Number Games, my 4th annual publication, shares several qualities with the previous 3. It's packed full of nerdy number trivia that I hope will blow your mind while showing you the beauty of mathematics and the world in which we live. Like my first book, *The Big Book of Numbers,* it walks you through the numbers 1 to 100. But what's different about this book is I've included more puzzles than ever before. When I say more, I mean hundreds more. And not just calculation exercises: *The Number Games* is loaded with logic puzzles, challenges and games to really stretch your grey matter.

I've also tried to make this book — in particular the answers — a 'teaching resource'. I hope that by the end of *The Number Games* you haven't just solved some of the questions I've posed, but that you've actually gained some insight into how to solve mathematical puzzles more generally. One of the joys of mathematics is the problem-solving skills it helps you develop. Logical thinking, strategic approaches and genius insights all play a role in solving the riddles in this book and will hopefully be part of your mental toolkit long after you've answered the final puzzle.

Some of the riddles are quite messy and are unlikely to reveal themselves to you instantly, so to avoid untidiness and mountains of scribble, I've loaded copies of the really gnarly brainbusters onto my website adamspencer.com. au. Feel free to print them off and make as much mess as you want before then creating a beautifully tidy and correct answer in your copy of the book.

And on the topic of mess, if you've borrowed this book from the library, best not to write in it at all!

I got great feedback on the puzzles from several students at Gosford High. Joshua Bito-On, Mathew Bowman, Theodore Brown, Georgina Buckmaster, Tamsin Caldwell, Liam Chaney, Donald Cheung, Sanjana Dinesh, Patrick Fang, Karl Harris, Eileen lee, Lucal Moncrieff, Amrut Nandakishar, Katerina Savkin, Joel Shortland, Connor Worth and Jenny Zhang — you guys rock! Thanks to super teachers Ben Watson and Cherylyn Rix for organising that.

Proofreading was provided by several young maths guns. Thanks to the thoroughly thorough Terry Shang, Jongmin Lim, Shane Arora, Jen Wakulicz, Michael Zhao and Daniel Collins from the University of Sydney, and Pauline O'Carolan, too.

The brilliant Sean Gardiner excelled himself this year and didn't just go through all the maths with a fine-toothed comb, he also invented some of the puzzles you see in the book. And if you think they're hard, you should have seen the original versions he provided that I had to 'dumb down'! The surfing scientist Ruben Meerman was again a great ally. Kjartan Poskitt seems, at least through email, to be an absolutely rocking dude. And Jon and Roy at Xoum as always made *The Number Games* look great. This book simply could not have happened without J & R being such stars.

Some of the puzzles are famous historic ones that I've reproduced verbatim. Others we've created just for this book. Some sit halfway between the two. I've endeavoured to include an acknowledgement any time a puzzle had a clear originator and sincerely apologise for any oversights in this regard.

So enjoy the book, digest the trivia, have a crack at the puzzles and let me know how it all goes. I'm at book@adamspencer.com.au on Twitter @adambspencer and Insta @adam_spencer1.

Welcome to ...
The Number Games.

Here we go!

001

Looking out for numero uno

If you've been given this book for Christmas, brace yourself for a beauty of a word to describe the very first, number 1 person or creature you meet upon walking out your front door on New Year's Day ... QUALTAGH!

Pronounced 'kwol-tog', it comes from the mid-19th-century Manx language which is spoken by a minority of the people who live on the Isle of Man, an island that lies between Great Britain and Ireland. Qualtagh translates as 'first foot', so it's sometimes used to refer to the first person to walk into your house on 1 January.

Whichever way you choose to define it, please, say hi to your qualtagh from me.

23	6		2	15
4		25		16
10	18			22
11			20	3
17	5		21	9

One fun

One of the most famous objects in mathematical puzzling is the 'magic square'.

A 5 × 5 magic square contains all the numbers from 1 to 25, once each, and each row, column, and main diagonal has the same sum.

When you've placed the missing numbers correctly in the magic square above, each row and column and the two long diagonals will add up to 65. I've given you some hints to start (after all, we're only at number 1!) so tell me this: where does the 1 go?

I know this book has only just started, but if you're feeling frisky, perhaps you can work this one out, too. Why do the rows of a 5 × 5 magic square like this always have to add up to 65?

One fun, two

To start us off, here's a typical counting problem to do with the number 1 that has confused high school students for decades.

I have a book with 100 pages, numbered, not surprisingly, 1 to 100. If I look at all 100 pages, how many times will I see the number 1 printed?

I'll get you started on this question for two reasons. Firstly, I'm just a decent sort of guy. But, more importantly, my solution will help you with an important part of the problem-solving techniques needed to answer a lot of questions in this book — in other words, you'll have a sensible strategy to make sure you get the correct answer.

If you just dive in here thinking, 'There's the 1 on the first page, then there's 11 which has two 1s, 100 starts with a 1 ... and of course there is 12 ... and there's also 21 ...' you'll quickly get lost.

Take a deep breath and realise this: the 1s in any number have to occur in some 'position' in that number. They can be at the end (we say these digits are in the 'units' column) or the next position along (the 'tens column'), or in the next position ('the hundreds'), and so on. To make sure we don't miss any 1s, let's go through the numbers 1 to 100 in this order.

Let's consider the ones that occur in the units column, that is in the numbers: 1, 11, 21, 31, 41, ... 81, 91; then the ones in the tens column of the numbers 10, 11, 12, 13, ... 18, 19; and finally, the one in the hundreds column of the number 100.

So we can see the number 1 will occur 10 + 10 + 1 = 21 times on the pages 1 to 100.

Here's the question. I'm reading a much bigger book now, with 1150 pages. By looking at all 4 positions that a 1 could occur on the pages of the book, how many 1s occur in the pages 1 to 1150?

It's cube-alicious

I have one large cube in front of me with sides all of 30 centimetres. I slice through the cube horizontally and vertically every 10 centimetres.

How many smaller cubes will I get?

Next I take a second large cube and before slicing it up I paint the outside yellow on all 6 faces. Now I slice it into smaller cubes as before.

How many of these smaller cubes will have only 1 yellow face?

Buccal phat!

How have your buccal fat pads been lately?

I know, you've got no idea what they are and didn't even know that you had some, right? Well, you've got two of them. Your buccal fat pads are the bits of a baby's face that give them adorable chubby cheeks. They play an important role in allowing babies to suck effectively when breastfeeding but in fact you keep them well into your life, only losing them in old age. It's the loss of the buccal fat pads that make some older people's faces appear 'hollowed out'.

When I searched for buccal fat pads on the internet, the first sites I came across were for cosmetic surgeons offering to help me reduce or remove mine.

If you're thinking of having that done, please think again. I'm sure your buccal pads make you look gorgeous.

Heads you win ... BIG

To celebrate the new year, your friend suggests a simple $1 bet. On 1 January you toss a coin. If it comes up heads, you win, but if it comes up tails, your friend pockets the dollar. Whoever wins can keep the cash or decide to come back on 2 January and go double or nothing.

If they win again they can try to double up again with another coin toss on 3 January that if successful would turn their $2 into $4 but, if they lose, the bet is over.

You decide, 'I'll play. But if I win, I'm going to go and go ... until I win $1,000,000!'

Let's say the coin comes up heads on 1 January. And keeps coming up heads every day. In which month would you win your $1,000,000? Even better, can you tell me the exact date?

Clique-bait

Academics who study our social networks often argue that our friends can be thought of as existing in circles around us. The very closest, our inner circle, are those 5 or so people we will go to in times of trouble — our closest friends. The next circle, our 'sympathy group' are people close enough that if they died or moved away forever we would miss them greatly. Then we move onto more casual acquaintances, and so on. It's thought that the total number of friends we can realistically manage is around 150 people.

And here's the kicker. According to Robin Dunbar, a professor of evolutionary anthropology at Oxford University, when we fall head-over-heels, rainbow-eyed, 'get a room, you two, I'm going to vomit', in love, our inner clique loses two people. It drops from 5 to 4, but one of those 4 is your new love interest, meaning that 2 family members or former BFFs have to take the fall.

Who knows, if that bestie of yours hates PDAs they may even be happy to step back a bit and join the sympathy group.

Seriously, you two ... get a room!

Liars, liars pants on fires!

One thing we'll be doing in this book is taking a walk down a beautiful place called Puzzle Place.

Puzzle Place is just like any street in any town, but perhaps with a slightly higher concentration of residents who *really* like mathematics and logic puzzles. Trust me, by the end of this book you'll have met some seriously geeky people on Puzzle Place.

One day, you are walking down Puzzle Place to visit your good friend Lionel Lyon. Out the front of Lionel's house you bump into the twin Lyon brothers.

It's ironic that they have this surname, because everyone knows that with the Lyon boys, one always tells the truth while one is always lying. What single yes/no question could you ask either of the Lyon boys to figure out if their dad is home?

Hint: you have to ask one of them a question that refers to the other one. Can you work out what that question would have to be?

Later that day, a little bit further down Puzzle Place, you see a brother and sister — the Fibbs kids — sitting at the bus stop. The Fibbs kids might lie or tell the truth at any given time, but at any given time, between them, at least one of them *always* tells a lie.

'I'm a boy,' says the kid with black hair.

'I'm a girl,' says the kid with red hair.

Does the brother or sister Fibbs have red hair, and how many of them are lying?

Three statisticians go out hunting together…

After a while, they see a rabbit. The first statistician takes aim and shoots exactly 5 metres to the left of the rabbit. At the same moment, the second statistician shoots exactly 5 metres to the right. The third statistician shouts out, 'We got him!'

This appalling joke relies on the fact that a statistician would take the average of 5 metres to the left and 5 metres to the right and get the result that the rabbit has been hit. Apologies for the quality of the joke, but brace yourself, there's more to come!

Flippin' switches!

Here's a classic, clever puzzle to limber you up. You're standing outside the door to a basement. The door is closed, so you can't see anything inside. Outside the door where you're standing, there are 3 light switches. Only one of them controls the light inside the basement. Your mate challenges you to figure out which switch controls the basement light, but she has a couple of rules:

- She tells you that the light in the basement is switched off to start.
- You can flip the switches any way you want, as many times as you want, but once you open the door to check, you can no longer touch the switches.

How do you figure out, without a doubt, which switch controls the light in the basement? The solution isn't about maths, but it's certainly clever. You can check it at the back of the book once you've had a crack.

HorseCow-SheepShip

*In April 1625, 3 ships set sail from Amsterdam in the
Netherlands to what we now know as America.*

*The ships were named Paert, Koe and Schaep, which is Dutch
for Horse, Cow and Sheep and you'll never guess what the cargo
was on each of them. Go on, have a go. That's right, Paert
carried horses, Koe carried cows and Schaep carried ...
yep, you get the drift.*

*I guess they were too busy trying to settle a new colony on the
other side of the world to bother dedicating too much time to
fancy names for their ships.*

*But Horse, Cow and Sheep certainly did their jobs well and
arrived in June of that year. And the 45 people they also carried
swelled the new population of what the settlers called The Man-
hatens (what we today consider Manhattan), to around 200.*

Sock it to me

There are two things that Dave really hates.

One is losing one sock in the wash and having a non-matching sock just floating around without a partner. The other thing he hates is having too many choices.

To make things easier, Dave threw out all of his old socks and bought just 3 types. Red socks with very cute blue mushrooms on them, green socks with a fetching black stripe around the top, and some yellow and blue polka dot socks that might sound a little bit adventurous but trust me, Dave nails it when he wears them. To avoid the problem of single socks, he went crazy and purchased 10 pairs of each of the socks, that's 20 socks of each colour.

Dave's wife Ming works late at night, so the one thing she really hates is getting woken up in the morning by a bright light as Dave gets ready for work. So every night he lays out his outfit in the living room and sneaks out to get dressed, leaving the bedroom, and his wife, in the dark.

One morning Dave wakes up, sneaks ever so quietly through the dark, gently opens the door, creeps into the living room, closes the door, turns on the living room light and, lo and behold ... no socks! He forgot to put them out last night.

He sneaks back into the pitch-black bedroom and, without turning the light on, grabs a handful of socks from the drawer. How many socks does he need to grab to make sure he has a pair?

And if he really wanted to wear the yellow and blue polka dot socks, how many socks would he need to grab to make sure he had a pair?

If you move more than 3 metres ...

... away from your car with the keys still in the ignition, if the cars is empty or if everyone still in the car is under 16, you've committed an offence in NSW, Victoria and Queensland.

And in a beautiful but bizarre distinction between the sort of people who might be on the footpath, if you splash mud on someone who is waiting for a bus, you're gonna pay for it, because you've infringed Rule 291-3 'Driver splash mud on bus passengers' — $180.

But splashing mud on other pedestrians ... that's fine!

Nurdle this

Perusing the fantastic twitter feed for the QI Elves @qikipedia, a place where you can learn some truly amazing and obscure stuff, I came across the word 'nurdling'.

According to the good people at the great BBC game show *QI*, nurdling means pushing old pennies into a hole in a bench (a game played in pubs), distinct from faffing about (doing nothing constructive).

But the more I looked into nurdling, the more interesting it became. Nurdling isn't just pushing coins. According to another website detailing a 2007 nurdling contest in rural England (where else?), a nurdle is a spherical object of not more than 3 ½ imperial inches and not less than 2 ½ imperial inches in diameter made of resilient material. Traditionally, flint pebbles were used. Nurdling involved pushing the stone down a road with a broomstick attached to a spoon (your grouting pole) while protecting yourself with a garbage bin lid.

Further, in the game of cricket, lately to 'nurdle' has come to mean to not hit the ball too hard, but just to ... to ... I guess to nurdle it away and take an easy run.

And finally, 'nurdles' are also small plastic pellets that can cause great pollution if loosed into our oceans. You could even say that I'm nurdling right now, in another sense of word meaning 'to gently waffle or muse on a subject which one clearly knows little about.'

So there you go, the world of nurdling is much more complicated than any of us imagined.

But back to the original nurdling that I read about. Just say I was nurdling 3 coins, marked A, B and C, into a hole in a bench. I might nurdle A into the hole first, followed by B, then C; or C first, then A, and finally B. How many different ways could I nurdle the 3 coins into the hole? Try to both calculate an answer and write down that many orderings of A, B and C.

Working this out logically so that you don't miss any combinations will introduce you to one of the most important mathematical tools when it comes to counting things.

Want some bonus points? Once you've worked out the number of ways 3 coins can be nurdled, if I told you I have some coins on a table that could be nurdled in 720 different ways ... how many coins would I have?

O's and X's i

Most of us have, at some stage, as kids, played a game of noughts and crosses.

You know how it goes — you start with a grid like the diagram above, and place Xs and Os until one of you gets three in a row.

To most people this is just a mild amusement for 10 year-olds. In fact, noughts and crosses is a numerical goldmine which introduces us to fascinating concepts in the mathematics of counting and game theory.

So, throughout the book we will take a quick look at some of the maths of the simple but gorgeous game.

It should be obvious that compared to a game like chess — where there are millions upon millions upon millions of possible move orders you can play for even a short game — with noughts and crosses there really aren't that many options.

In fact, again as opposed to chess, good ol' Os and Xs is *so* simple that we've been able to crunch out all the possible ways any game can go.

Let's assume that X always goes first. Also, if you have two boards with Xs and Os on them, but one is just a rotation or reflection of the other, let's say they are in the same position.

It turns out there are only 138 possible positions at the end of a game. Of these 138 possible 'end positions', 91 distinct positions are won by X, 44 distinct positions are won by O, and there are 3 distinct possible drawn positions.

Here's the question: can you find the 3 possible draws?

Three little boxes

Joseph Louis François Bertrand was a French mathematician who lived in the 19th century.

Among other things, he gave us this classic logic puzzle called 'Bertrand's Box Paradox' from his 1889 ripping read *Calcul des probabilités*. Here goes.

You have three unmarked boxes before you:

- One box contains two gold coins,
- Another box contains two silver coins, and
- The third box contains one gold coin and one silver coin.

You close your eyes, reach into one box and draw out a gold coin. What are the odds that if you draw out the other coin, it will also be gold?

Now, let's change it up a bit. This time you have 3 boxes:

- One labelled 'gold',
- One labelled 'rocks', and
- One labelled 'gold and rocks'.

Not only that, but all the labels are incorrect.

I'll let you pick one item, blindfolded, from the box of your choice, then you can choose to leave with any box. How do you make sure you leave with the box full of only gold?

The cat's whiskers

Graphene is an amazing substance. It's made out of carbon atoms arranged in a hexagonal lattice that stretches out in two dimensions ... but is only one carbon atom thick. In case you're wondering, 'one carbon atom thick' isn't very thick at all. In fact, it's 0.3 nanometres thick. Or, to put it another way, something like 400,000 sheets of graphene, lying on top of each other, would be about as thick as this sheet of paper this sheet of paper! Because of this, graphene is considered for all intents and purposes to be a two-dimensional substance.

Despite being so thin, graphene has incredible properties. It is almost completely see-through, yet so dense that not even the smallest gas atom, helium, can pass through it. It's 200 times stronger than steel, but 6 times lighter. And it also conducts electricity and heat very well.

One of the coolest explanations I've read about the strength of graphene comes from the committee that awarded the 2010 Nobel Prize for Physics to Andre Geim and Konstantin Novoselov from the University of Manchester 'for groundbreaking experiments regarding the two-dimensional material graphene'. If you made a hammock of graphene that was 1 metre by 1 metre, you could rest a 4 kilogram cat on that hammock. But here's the really cool part: the hammock itself would weigh less than one of the cat's whiskers.

So where are all the graphene cars and aeroplanes? And where's my graphene mobile phone that I can roll up in a ball and stick behind my ear? Well, we still don't know how to produce graphene in industrial quantities. Nevertheless, if you have ever drawn on a piece of paper with a lead pencil, you have most likely made graphene!

True or fourlse?

Here are four statements.

1. The number of false statements here is one.
2. The number of false statements here is two.
3. The number of false statements here is three.
4. The number of false statements here is four.

Which of the above statements, if any, is true?

Remember the 4 Ps

It sounds like some bizarre modern-day action film, but in Tasmania people are being urged to use the '4 Ps' to stop the scourge of giant fatbergs.

Okay, let's take this one step at a time. What are the 4 Ps?

The 4 Ps are Pee, Poo, Puke and Paper.

These are the only things that should be flushed down a toilet system. The first three are obvious, but it's the fourth one, paper, that is the kicker here. Good old fashioned toilet paper breaks down in water and can be easily handled by the sewer system. But more complex substances including baby wipes do not break down and clog the pipes.

The clogging can get so bad that the accumulated wipes, along with food fats and other things that aren't in the 4 Ps, can glob together forming something like an iceberg of fat. That's right … the dreaded fatberg.

It sounds like a bit of fun until you realise that even a small sewer system like the one in Launceston, Tasmania, has to remove one whole TONNE of wipes per fortnight from its machinery.

One of the worst fatbergs ever encountered was in 2013 in London when a 10-tonne monstrosity, reported to be the size of a bus, was removed from the pipes.

Flag Fun

Time to continue our stroll down Puzzle Place.

Number 4 Puzzle Place is the daycare centre. And today the kids are making flags. The flags are a basic design and feature just 3 vertical stripes. The kids can choose from 4 colours — Red, White, Blue and Green. How many different types of flags can they make if:

1. You can have neighbouring stripes the same colour, even all three stripes the same so, say, just an entirely green flag?
2. You can't have a colour repeated anywhere on a flag?
3. You must have a repeated colour on the flag?
4. The end stripes can be the same colour or different, but neither end stripe can match the centre stripe?
5. You can't have any colours repeated anywhere, but instead of a flag you're making a 'placemat' for a table? So Red - White - Green is the same as Green - White - Red because you can spin the mat around on the table and so these two mats actually look the same.

The answers to these questions introduce you to some very important maths tools that we use for counting, so make sure you check them out and understand them. They will come up again at later stages in the book.

Humans don't pass chimps in intelligence until we are about 4 years of age

University of Queensland researchers created an experiment that required planning to successfully catch an object being dropped down one of two tubes.

While 2 and 3 year olds mimicked apes and placed both hands over just one tube and hoped for the best, by the age of 4, children had worked out it's best to place one hand over each tube, thus covering all possible outcomes.

The researchers claim this 'foresight' is one of the defining features of intelligence that separates us from animals ... and kicks in around 4 years of age.

Winner, winner chicken dinner!

Three travellers finally get to their hotel really late. As they check in, exhausted, they ask the kitchen if they could get some food. 'We've only got chicken nuggets,' says the manager. 'No problem,' they reply. 'Fry up every one of them you've got and bring them to our room ... pronto, please!'

The travellers get to their room, but they are so tired they collapse asleep in their beds before the food arrives. The room service guy, exhausted and hungry, can't resist sneaking a cheeky half before laying them at the door.

Fifteen minutes later, one of the travellers wakes up to the smell of chicken nuggets. Our traveller does the right thing and only eats one-third of the nuggets then promptly falls back asleep.

A few minutes later, another traveller wakes up and, assuming the manager let himself in and that this is all the nuggets, helps himself to one-third of the nuggets on the tray and ... promptly falls back asleep.

The final traveller wakes up shortly after and, you'll never guess, that's right, assuming she's the first to wake up, eats a third of the nuggets and falls back asleep.

In the morning there are 4 stone cold nuggets still on the plate.

How many nuggets did the manager entrust to the room service guy to take up to the room?

The first high five!

On 2 October 1977, the Los Angeles Dodgers faced off against the Houston Astros at LA's famous Dodger Stadium. The Dodgers lost 6-3 and as they trooped out of the stadium disappointed in defeat, little did the 46,000 fans know they had witnessed a piece of sporting history.

It all happened in the bottom of the sixth innings. If you've never understood what that meant, it means that the Astros had already had their 6th turn to bat and it was now the Dodgers having a go. In a tremendous innings, LA hit three home runs to bring the score from 0-2 up to a 3-2 lead. People were obviously pumped.

But it was more than just a good innings. In hitting the second homer of the 6th, Dusty Baker brought up 30 home runs for the season and made the Dodgers the first team in history to have 4 players reach that mark in a single season. Dusty's great friend on the team, Glenn Burke, a young outfielder and phenomenal athlete, waited for Baker as he came toward the home plate, waving his hand in the air in celebration. Not sure what to do, and caught up in the moment of locking the scores at 2-2, Dusty Baker leapt in the air and slapped his hand. 'His hand was up in the air, and he was arching way back,' says Baker, now 62 and managing the Reds, '... so I reached up and hit his hand. It seemed like the thing to do.'

It might all have ended there except that moments later Glenn Burke smoked one over the fence, thus hitting his first ever major league home run. As he reached the home plate, Dusty Baker returned the high-slapping favour and the high five was born!

Tragically, the game was not televised so no footage remains of this, the first ever sporting high five.

Think you're smarter than a 10 year old?

This cracking problem is taken from the 2016 International Singapore Mathematics Competition.

It's an awesome initiative to get primary-aged schoolkids into maths and problem solving. Having said that, to be honest, a lot of these questions aimed at 10 year-olds will get adults thinking pretty hard! Check out their website at www.ismc.sg when you've got a spare moment.

Sally was given a set of 5 cards numbered 1 to 5, and Peter was also given a set of 5 cards numbered 1 to 5. They were then blindfolded and told to pick a card from their respective sets.

The sum of the numbers from the two cards was told only to Sally and the product of the numbers was told only to Peter. They were then told to name the numbers on the two cards they had chosen. Below is what each of them said.

Peter: I don't know the two numbers.

Sally: Now I know the two numbers.

Peter: I still don't know the two numbers.

Sally: Let me help you. The number I was just told is larger than the number you were just told.

Peter: Now I know the two numbers!

So, what were the two numbers? Here's a hint if you're stuck (but if you're all over it, stop reading now!) Peter has been told the product of the two numbers. But he doesn't know which numbers give him that product. What does that tell us about the product?

High, er, five?

It's not just American baseballers and overly zealous parents coocooing with their kids at the local shops who can dispense a pretty mean high five.

A team of Austrian veterinarian scientists from the University of Vienna, led by the awesomely-named Raoul Schwing, ventured to a mountain on the south island of New Zealand where they found a population of wild kea parrots. They played 5-minute recordings of different sounds to the birds, including a recording of the kea's distinctive laughter-like warble. When the parrots heard the warble they went crazy. They warbled back, performed aerial acrobatics and exchanged high fives with their claws. When the 'laughter' stopped, they calmed down and went back to whatever they were doing. None of the other sounds got such an energetic reaction from the parrots, only the warble.

So it seems humans don't have a mortgage on high fives — keas love them too.

Strictly speaking, a kea's claw has 3 digits so we should say that they dispense 'high threes', but for me, high five sounds just that little bit cooler. Do we agree? C'mon … don't leave me hangin'.

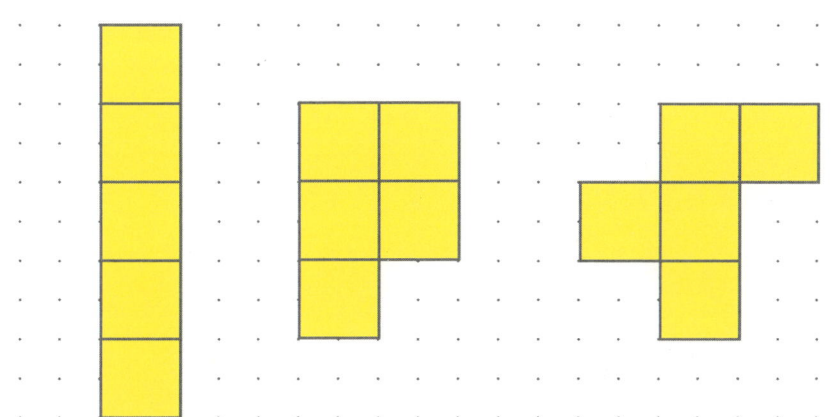

Hey ho, pentomino

A 'polyomino' is a shape (or polygon) that is made by joining a number of squares edge to edge.

When we use 5 squares we call the polyominoes 'pentominoes' from the Ancient Greek *pénte*, meaning 'five'.

Above are 3 pentominoes. Now, the pentomino below ...

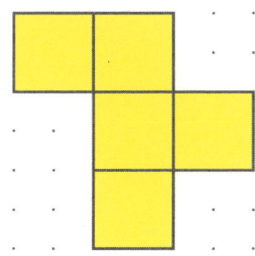

... is just the previous pentomino reflected.

See if you can come up with all 12 'free' pentominoes. Make sure you don't accidentally include 2 pentominoes on your list that are rotations or reflections of each other.

The world's smallest tree grows to only about 6 centimetres tall

It goes by the name Salix Herbacea, or dwarf willow, and is found on the tundra of Greenland.

On point

If I want to draw a path that connects the 4 points of a square, touching each point only once and finishing back where I started, there are two ways I can do this:

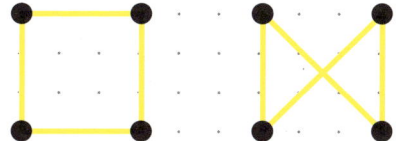

Any other paths are just one of these two reflected or rotated.

For the 5 points of a pentagon, there are 4 ways I can do this:

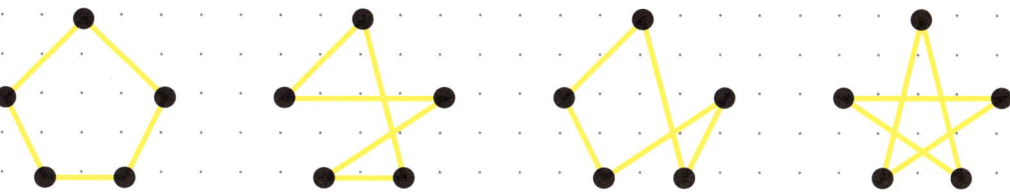

Again, any others you can find are just reflections or rotations of one of these 4 paths.

Now, for the 6 points of a hexagon, there are 12 different paths. Can you find them?

If you're wondering what my favourite Russian nursery rhyme is ...

Look no further than this little gem:

Grandpa found a pineapple in the field.
He didn't think that it was a hand grenade.
He pulled out a knife and got ready to eat,
They found his ass six kilometres away.

Of course, it's much more poetic in the original Cyrilic.
But if you'd like to read it as a Russian would,
the phonetic pronunciation is as follows:

Dedushka v pole nashel ananas,
On ne podumal, chto eto fugas.
Nozhik dostal on, sobralsia poyest',
Zhopu nashli kilometrov za shest'.

Six shakes six shakes six shakes

Here's the scenario. Six people catch up for lunch. They do the polite thing and shake each other's hands before sitting down. How many handshakes occur?

Have a bit of a think about it, but if you need a hint, here goes.

Seriously, have a crack at it first. Still stuck? Okay, here goes ... for real this time.

If you just let them all randomly shake hands and try to keep track you will never nail this down.

Can you think of a neat, organised way they can shake hands that is easy to count and covers every possible handshake just once?

Bonus fact for you: people who have 6 digits on each hand (or feet) have 'hexadactyly'. There you go!

Dice, dice, baby ...

Later in this book you'll meet some truly mind-bending dice. You know: fun, but a bit freaky. But that's for another time. Let's warm up by playing some games with standard 6-sided dice.

We'll start easy and get harder.

Here goes! If:

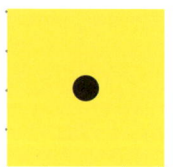 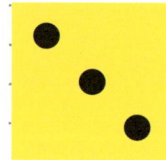 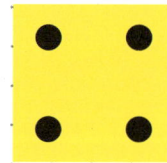 = 8

And:

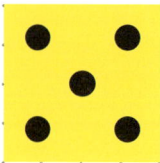 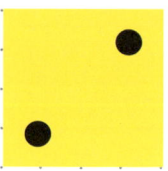 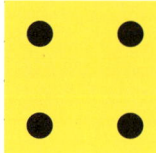 = 11

... What is a roll of:

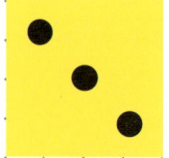 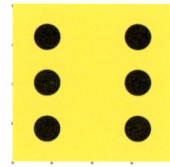  = ?

... worth?

Great!

Okay, now let's go medium difficulty.

If a roll of:

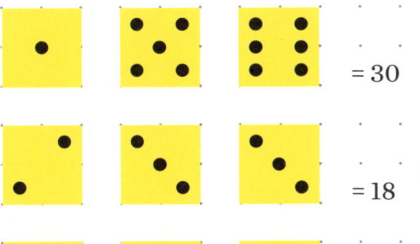

= 30

= 18

= 125

which roll earns more out of the following two?

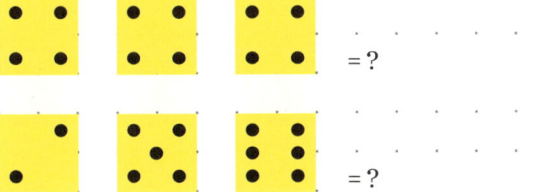

= ?

= ?

And a toughie for you. If:

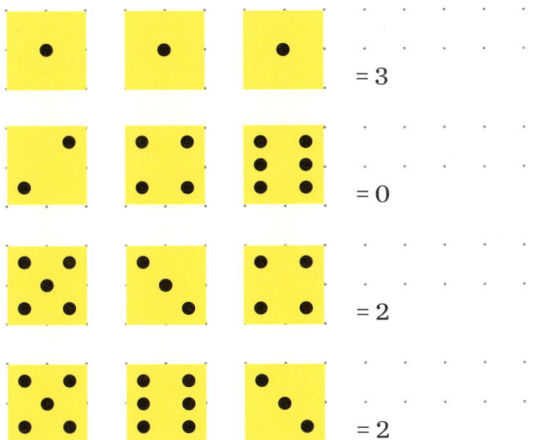

= 3

= 0

= 2

= 2

... how much does this get us?

= ?

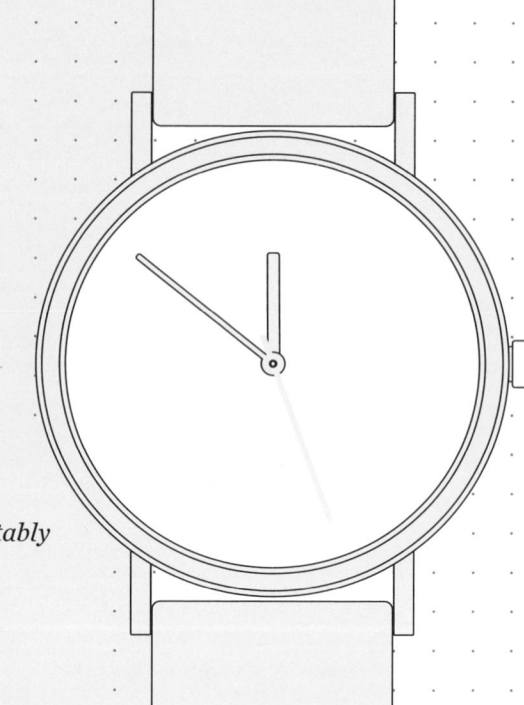

Got a sec?

*You might have heard of the 'Doomsday clock'
— a symbolic gauge of our chances of killing
ourselves through global catastrophe (most notably
nuclear war or, more recently, climate change).
It was inaugurated in 1947 by the* Bulletin of
Atomic Scientists, *which was itself founded by
members of the Manhattan Project (the R&D
undertaken during WWII to develop the first nuclear
weapons). So it's fair to say this isn't your average,
tin-foil hat-wearing conclave of garden variety loons
complete with sandwich boards.*

Each year, the Bulletin *gather a group of luminaries
(including, at the moment, 18 Nobel laureates)
to assess the risks and update the clock. At its
inception, the clock was set at 7 minutes to
midnight. Within 6 years it had lost 5 minutes,
as the US and Soviet Union tested thermonuclear
devices months apart. Fortunately, over the coming
years, it gained time. In fact, despite some ups and
downs, by 1991, it was 17 minutes from midnight, with
the signing of the Strategic Arms Reduction
Treaty, and the dissolution of the Soviet Union.*

*Bad news, though. Save for a slight gain in 2010,
it's been nearing midnight ever since. Worse news:
the latest finding by the committee, on 26 January
2017, saw the clock lose even more time: we're
now at two-and-a-half-minutes to midnight.*

That's a pretty important 150 seconds.

Seven points!

Above is a glorious page in AFL history. It's a match between New South Wales and Queensland, held in the 1933 Australian Football Carnival at the SCG. And below? Why, it's another mathematical mind-bender to wrap your head around!

In a game of 'toofball', a team scores 7 points for a yrt and 3 points for a laog (hey, it's a new type of footy, okay?). Teams score yrts and laogs independently of each other.

Obviously it's impossible for a team to end up on 2 points, or 4, or 5. You should also be able to see that it's impossible to score exactly 8 points.

What's the largest score a team cannot end on in an infinitely long game of toofball?

You don't have to wait indefinitely for the answer if you're stuck. It's at the back of the book. But, for a bonus point, if yrts were worth 11 and the largest unreachable score was 39, what's a laog worth?

TYPE 1: Separate hard lumps (hard to pass)

TYPE 2: Sausage-shaped but lumpy

TYPE 3: Sausage-like, with cracks on its surface

TYPE 4: Sausage or snake-like, smooth and soft

TYPE 5: Soft blobs with clear edges (passed easily)

TYPE 6: Fluffy pieces with ragged edges. Mushy.

TYPE 7: Entirely liquid, with no solid pieces

Stool school

From being the impresario of intestinal integrity to the devil of digestive disaster, here are the 7 different ways you can formulate your faecal firmness or process your poo's passability. I'm referring, of course, to the 7 types of number 2s that can be found on the Bristol Stool Chart.

In case you're wondering, you're ideally looking for 3s and 4s. Bear in mind, though, that we've taken a fair bit of liberty with our diagram above. Google for more, er, accurate images. But one thing I can guarantee: if your poo is the colour of any of the ones above, see your doctor.

A fine vintage (puzzle)

Here's a corker from Henry Dudeney's awesome *Canterbury Puzzles and Other Curious Problems.*

It was first published in 1907 and features a series of nifty puzzles based on Chaucer's famous characters in his *Canterbury Tales*.

Strap your thinking cap on.

Abbot Francis sends for his cellarman and complains that a particular bottling of wine is not to his taste. He asks how many bottles he had produced. The cellarman tells him that there had been 12 large and 12 small bottles, and that 5 of each have been drunk. The abbot replies that 3 men are waiting at the gate, and orders the cellarman to give each of them some combination of full and empty bottles so that each man receives the same quantity of wine and combination of bottles.

How can the cellarman do this? He has 7 large and 7 small bottles full of wine, and 5 large and 5 small bottles that are empty. A large bottle holds twice as much wine as a small one, but a large bottle when empty is not worth two small ones — hence the abbot's order that each man must take away the same number of bottles of each size.

Note that you can't pour between bottles, and you want to get rid of all 12 bottles.

Answer, of course, at the back of the book.

'Zenzizenzizenzic'

Arguably the coolest mathematical word of all is zenzizenzizenzic.

This single-word tongue twister is an old-fashioned way of expressing the eighth power of a number. We can calculate 2^8 which is $2 \times 2 \times 2 \times 2 \times 2 \times 2 \times 2 \times 2 = 256$ and say '256 is the zenzizenzizenzic of 2'.

Seriously, you can say that! Try to work it into conversation sometime.

It's thought that the first person to use this awesome term was the equally awesomely named Welshman Robert Recorde who in his 1557 ripsnorting read The Whetstone of Witte explained that the word he spelt zenzizenzizenzike 'doeth represent the square of squares squaredly'.

If you think you have a contender for a COOLER maths word, hit me up and let me know.

Alternative fact(orials)

Have a go at this if you'd like to check if your arithmetic skills are up to scratch.

A factorial is the product of a whole number, and all the whole numbers below it, down to one — for instance 3! (how we write 3 'factorial') is $3 \times 2 \times 1 = 6$.

Well, it turns out that 213 is the smallest 3-digit number whose square is the sum of distinct factorials. Who knew? Certainly not me until I read it, and I'm sure you won't be offended to know that I assume you didn't know it either!

So what does it mean?

It means that there are a series of factorials that when added together give $213^2 = 45,369$.

If I told you that there were 5 factorials you needed to add together and, further, that 8! (I'm not just excited about this, remember that we write '8 factorial' as '8!') was the largest of the factorials involved, can you find the other 4 factorials needed to give us the sum?

Of course you can ... off you go.

In April 2017 an 8-year-old boy went to his local Ohio McDonald's

So what, you say? Well, he drove — himself — in his parents' van, having taught himself to drive by watching YouTube videos.

How about smarter than an eighth grader?

The prestigious Mathcounts National Competition is held every year in the United States, and is best thought of as the maths version of a national spelling bee.

Two hundred or so completely awesome young guns are brought together to answer tougher and tougher questions until the top 12 face each other in a speed maths battle to the death.

The final question in Mathcounts 2012 was this little ripper:

A bag of coins contains only 1c, 5c and 10c with at least 5 of each. How many different combined values are possible if 5 coins are selected at random?

Indiana eighth-grader Chad Qian tore this apart in just seconds. I expect it may take you a fair bit longer, but think logically and it is very doable.

Fancy a bonus question? You got it.

The highest total you could get is obviously 50c. What is the longest string of consecutive amounts that you can get and what is the longest string of consecutive amounts below 50c that you can't get when choosing exactly 5 coins?

If you liked this there are a few more Mathcount classics throughout *The Number Games*.

It takes about 9 hours to read Amazon Kindle's T&Cs aloud

We've all had a giggle at the 'terms and conditions' that flash up on the screen when you upgrade some software on your phone or purchase a new app. But even by the standards of online terms and conditions, the Amazon Kindle legals are absurd. Consumer advocate Choice *took a look at the Kindle T&Cs and calculated that the document would take almost 9 hours to read.*

The Kindle contract is a whopping 73,198 words long. In case you're wondering, that's longer than Shakespeare's Romeo and Juliet, Hamlet *and* Macbeth ... *put together!*

In fact, in one of the funnier online postings I've seen in a while, Choice *even hired an actor, Laurence Rosier Staines, to read the Kindle terms and Conditions, in full ... and filmed him doing it. They posted 9 one-hour-long instalments on YouTube in case you've ever got a massive amount of spare time on your hands.*

Laurence, well read, my friend.

A square affair

Place the numbers 1 to 9 in each of the squares above so that each row, column and main diagonal sums to the same amount.

Want a hint? Once you've put all 9 numbers in, what must each row add up to if they all equal the same?

And what number must logically go in the middle of the grid?

Give it a go!

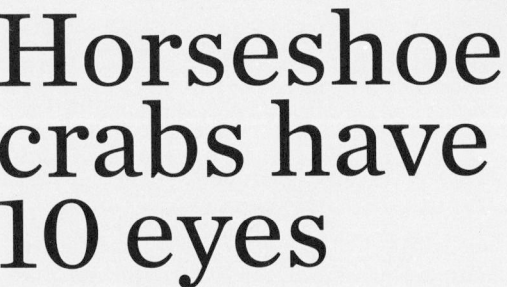

Horseshoe crabs have 10 eyes

They've also been around over 450 million years, have an average heartbeat of 32 per minute (fit!) involving their long tube of a heart that extends almost the length of their body, shed their shells up to 17 times in their lifetimes, and have two compound lateral eyes on top of their shells that do most of the serious looking around for other crabs to get friendly with.

The crabs can also sense light through less complex lateral eyes, two median and one endoparietal eye, all on top of their shells. Count the two eyes on the bottom of the shell near the mouth, and the series of light receptors in the tail and you have 10 horseshoe crab eyes altogether. They can do everything from help the crab position itself when swimming, to detect ultraviolet light.

Lookin' good!

Still think you're smarter than a 10 year old?

O kay, so you warmed up back at number 5. Feeling pretty good? Here's another from our Singaporean friends. This one's a little harder, but have a go.

There are 4 keys and 4 locks. What is the maximum number of times you need to try the locks so as to match all 4 keys to their locks?

A. 4
B. 6
C. 10
D. 16

The answer, as you'll find with all of these puzzles ... is at the back of this very book.

Spit the bishop

If, like me, you love sports, you will know one of the gravest accusations that can be levelled against any competitor is that of 'tanking'.

Tanking refers to not trying to win. In some cases, tanking involves taking an 'I just couldn't be bothered' attitude into a contest, but, in its worst form, can cover players deliberately trying to lose.

Australian tennis's frustrating boy genius Nick Kyrgios was booed off the court for his effort at the 2016 Shanghai Masters when he seemed to be trying to lose as quickly as possible. At one stage, he lobbed a soft serve to his opponent and walked away before it had even been returned. Another time, he asked the umpire to please call time on the game early so he could go home to bed.

But while accusations of tanking normally centre around brattish tennis players and football teams trying to finish lower on the ladder to get draft concessions for the next season, even the noble and dignified world of chess is not immune.

The 2017 Gibraltar Masters was one of the world's most prestigious chess tournaments featuring several of the world's top 10 players. However, a lot of the media coverage was taken away from the likes of Fabiano Caruana, Hikaru Nakamura and Veselin Topalov when the world's number one female player, Hou Yifan of China, cracked the angries in her last game.

You see, Hou loves to play against the best men in the world, but during the Masters, the random computer pairing program pitted her against female opponents on 7 out of her first 9 games. By the time she'd reached the final round, she was over it — and even though she was drawn to play against the Indian man Babu Lalith, she deliberately threw the game. After only 10 moves!

For those of you who don't understand chess notation, I go into detail in my earlier book World of Numbers *(PLUG!). For those that do, I can tell you Hou played the following suicidal moves:*
(1) g4 d5 (2) f3 e5 (even at this stage Lalith would have been thinking, 'What on Earth is going on here!') (3) d3 Qh4+ (4) Kd2 h5 (5) h3 hxg4

And with that, with only 10 moves having been made, Hou resigned. In doing so, she suffered the quickest loss in Grandmaster Chess history. She later apologised, but I say 'good on you, Hou – we've all been there!'.

It was reported that at this competition, the ratio of women to men was 1:4. So just what were the chances that Hou would have 7 out of her first 9 opponents be women?

Let's break the problem into smaller chunks. What are the chances Hou's draft started: FFFFFFFMM? The chance for each female opponent draw was $\frac{1}{5}$, and the chance for each male opponent draw was $\frac{4}{5}$. So we're multiplying $\frac{1}{5}$ by itself 7 times, and $\frac{4}{5}$ by itself twice. That means there was a $(\frac{1}{5})^7 \times (\frac{4}{5})^2 = 0.0008192\%$ chance of drawing FFFFFFFMM. But of course, the order shouldn't matter, so drawing FFFFFFFMM is as likely as drawing FFMFFFMFF, say. Just how many lists of seven F's and two M's are there?

Well, imagine starting with a string of nine F's and turning two of them into M's. The first M can take any of 9 places, and the second M can take any of the remaining 8 places. This would seem to give $9 \times 8 = 72$ possibilities, but since choosing position 1 first and position 2 second is the same as choosing position 2 first and position 1 second, we've actually double-counted, meaning there are exactly $9 \times \frac{8}{2} = 36$ such arrangements.

So the chances of drawing 7 females in your first 9 rounds is $36 \times 0.0008192\% = 0.0294912\%$, or about three-hundredths of a percent. As you might expect, this approximation doesn't even change if you throw in the chances of drawing 8 or 9 female opponents, so 0.03% is also the chances of drawing AT LEAST 7 women. In fact, the chances of any draft of 9 rounds having more than half the opponents women is a measly 1.958144%.

No wonder Hou was miffed!

Apollo 11 launched from Kennedy Space Centre on 16 July 1969

It carried a precious payload: the lunar module Eagle, *along with astronauts Commander Neil Armstrong, Command Module Pilot Michael Collins and Lunar Module Pilot Edwin 'Buzz' Aldrin.*

Four days later, in front of an estimated 530 million terrestrial TV viewers, Armstrong would make history with 'one small step for a man, one giant leap for mankind', as he stepped off the Eagle's ladder and onto the Moon. Aldrin joined him about 20 minutes later. Together they spent 21 hours, 36 minutes on the Moon, although their EVA (that's 'Extravehicular Activity' — the bit when they were out-side the module for us Earth-lubbers) lasted little more than two-and-a-half hours.

8								
		3	6					
	7			9		2		
	5				7			
				4	5	7		
			1				3	
		1					6	8
		8	5				1	
		9				4		

This one goes to 11

So you like your Sudokus, do you? Reckon you're pretty decent at them? Well, let me introduce you to Finnish mathematician Arto Inkala. In 2012, Inkala devised what he claimed to be the 'World's Hardest Sudoku'.

And it's true: this little gem is fiendishly difficult.

According to one newspaper, if you regularly tackle 5-star difficulty level Sudokus ... this one's an 11.

Good luck. Answers at the back of the book.

Heads or tails, you win

Flip a coin. Any coin will do.

Now, go ahead and enter your result (TAIL or HEAD) as the answer to 1 across in the crossword grid opposite.

Mathematician Jerry Farrell of Butler University devised this little number which works despite the outcome of the toss.

A 3 × 4 crossword? Phht. Easy, you say! Oh yeah?

I'll be honest, Jerry's clues are pretty hard.

Take a look at his original (really tough) clues on the opposite page and have a go.

But, for mere crossword mortals, I've also provided my own set of slightly easier clues below.

Good luck!

Across

1. The result of your coin toss
5. The Earth Goddess in Wagner's opera *The Ring of the Nibelung* (go ahead — Google it)
6. Looked at

Down

1. Half of a word that describes the sound of laughter
2. Three letters that can be put on the end of 'station' to spell a new word
3. Three letters that can be put on the end of 'dec' to spell a new word
4. A type of male human

1	2	3	4
5			
6			

Across

1. The result of your coin toss
5. Wagner's Earth goddess
6. Word with one or green

Down

1. Half a laugh
2. Station terminus?
3. Dec follower?
4. Certain male

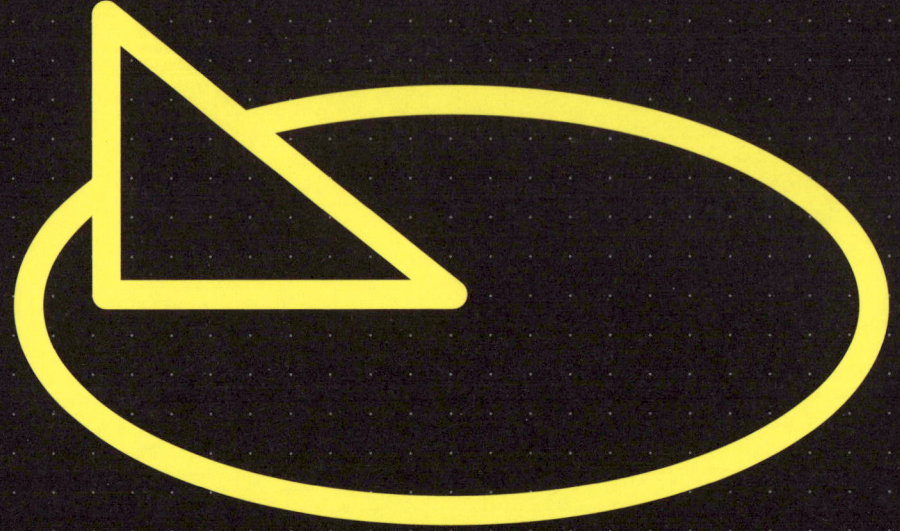

12 noon was originally 3 pm!?!

The word 'noon' comes from the Latin phrase nona hora, *which means 'the ninth hour of the day'. Back in the monasteries of Rome and Western Europe in the Middle Ages, you were up and at them at 6am, so the ninth hour began at 3pm.*

By the 12th century, noon was on the march and began to mean 'midday' or 12pm local time. By the 14th century, noon no longer meant 3pm.

As easy as 123...456

Here's a question from the 2016 United Kingdom Junior Maths Challenge.

About 4 in every 5 of the 11 to 13 year olds who took the test got this right ... so no pressure!

Which of the following statements is false?

1. *12 is a multiple of 2*
2. *123 is a multiple of 3*
3. *1234 is a multiple of 4*
4. *12345 is a multiple of 5*
5. *123456 is a multiple of 6*

All mammals take about 12 seconds to empty their bowels

Apologies if you're still reeling from the Bristol Stool Chart which we introduced back at 7, but here's another number two truth. Almost all mammals, large or small, weighing from 4 kilos to 4000 kilos, take about 12 seconds to empty their bowels, commonly creating two pooey pieces in doing so. Patricia Yang (a Mechanical Engineer PhD Student of all things) at Georgia Institute of Technology, Atlanta, shared this discovery in a 2017 edition of the journal Soft Matter.

Another fascinating faecal fact is this: one of the crucial parts of your body in helping you do the do is a layer of mucus in your colon. This helps the poo slip out easily. If the mucus is absorbed by your waste you get constipated and it gets a lot harder to say goodbye to the goo. And if you didn't have this layer of mucus and applied no pressure at all, you'd only empty your bowels once every 500 days. Patricia warns, 'It would be shortened to 6 hours if you apply maximum pressure, but I believe you'd still need to see a doctor.'

Option A

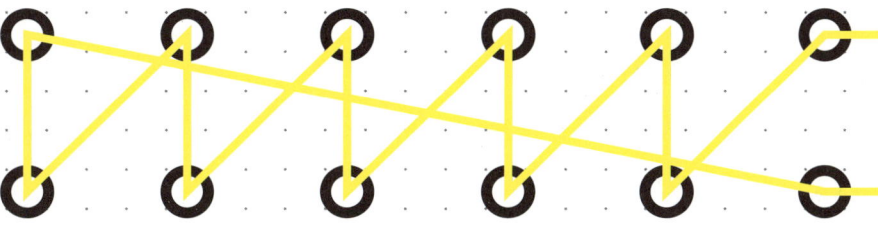

Option B

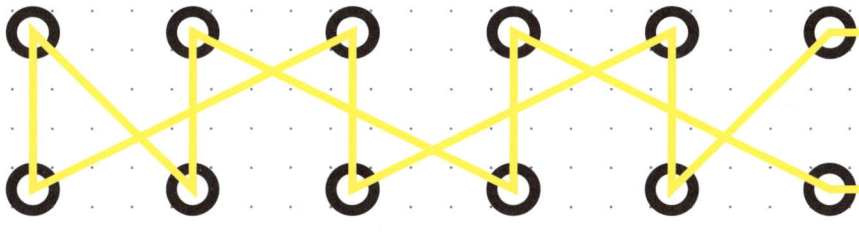

Option C

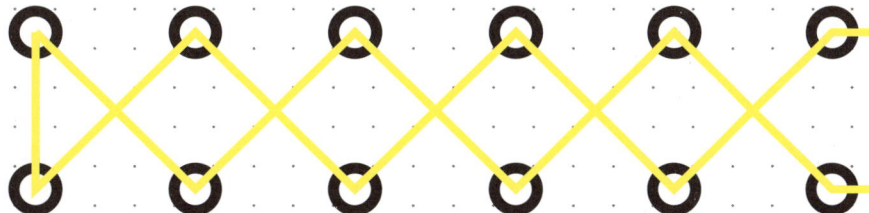

It's a tie!

O r is it?
Without measuring with a ruler (that is, just by looking with your eyes), arrange the ways of lacing a shoe above in order from the way that uses the least lace up to the way that uses the longest lace.

013

It cost $13 to stop a computer virus taking over the world

Okay, my language is a little bit flowery here, but the WannaCry virus was a piece of ransomware that tore across the IT of the world in May of 2017. Like all good ransomware, it disabled your computer and encrypted your files, providing a message that it would only undo the damage if you paid $300 in the virtual currency Bitcoin.

But WannaCry was regularly checking back in to a domain name to get permission to keep spreading. The domain name had probably been put into the program to allow the creators to use it as a 'kill switch' some time down the line. A 22-year-old British cybersecurity researcher noticed this and registered the domain name. Surprisingly, this stopped WannaCry in its tracks ... at least for now.

Registering that domain cost the princely sum of $13.

Lucky for some ...

The 13 prisoners stood in a line as the evil warlord spoke.

'I have captured 13 of you, but only have room to take 8 of you onto my ship. Five of you will be lucky enough to stay on this island and fend for yourselves. Speaking of 5, this is what I'll do. Stand in a circle and number off from 1 to 5 in the direction the hands of a clock turn. Each time one of you says a 5, you must board the ship. The next person in the circle starts at 1 again until we hit 5 and that person boards the ship. Then the next person in the circle must start counting from 1 again. We will do this until I have filled my ship with 8 of you and the remaining 5 can go free. Form a circle ... NOW!'

You hurry into a circle with your best friend standing immediately to your right. The warlord looks at you and says, 'You look as though you like a challenge. You can choose to start the counting yourself, or pick someone else to start counting. Where shall we start? ... NOW!'

'Gee, this guy likes to shout the word 'now',' you think as your mind races through the possibilities. Should you start the count? If not, how many people away from you will you choose as the starting point to save both you and your friend?

Your time starts ... you guessed it ... NOW!

De Fib

One of the most famous lists of numbers in all of mathematics is the Fibonacci numbers.

It is an example of what mathematicians call a 'sequence' and is generated by a simple rule. Starting with 0 and 1 we simply add the last two numbers of the sequences to generate the next number. So 0 + 1 = 1 and our sequence now starts 0, 1, 1. Using 1 + 1 = 2 the sequence is now 0, 1, 1, 2 and by continuing like this you soon get 0, 1, 1, 2, 3, 5, 8, 13, 21, 34, 55, 89, …

The Fibonacci Sequence has all sorts of amazing properties. If you add up the first n terms of the sequence you get the $(n+2)$th Fibonacci number minus 1.

We write this snazzy fact as:

$$\sum_{i=1}^{n} F_i = F_{n+2} - 1$$

Don't panic about the symbols. It just means for $n = 8$ we get the sum of the first 8 Fibonacci numbers: $0 + 1 + 1 + 2 + 3 + 5 + 8 + 13 = 33$ and the 10th Fibonacci number is 34, so yes the first sum does equal $34 - 1$.

The sequence can be found in another famous mathematical table called Pascal's Triangle. In this triangle the first row is a 1, the second row is two 1s, and then you will notice that each new row begins and ends with a 1 and the other numbers in the row are given by the sum of the two numbers directly above it.

If we add up along the 'shallow diagonals' of Pascal's Triangle as shown opposite, we get … the Fibonacci Numbers.

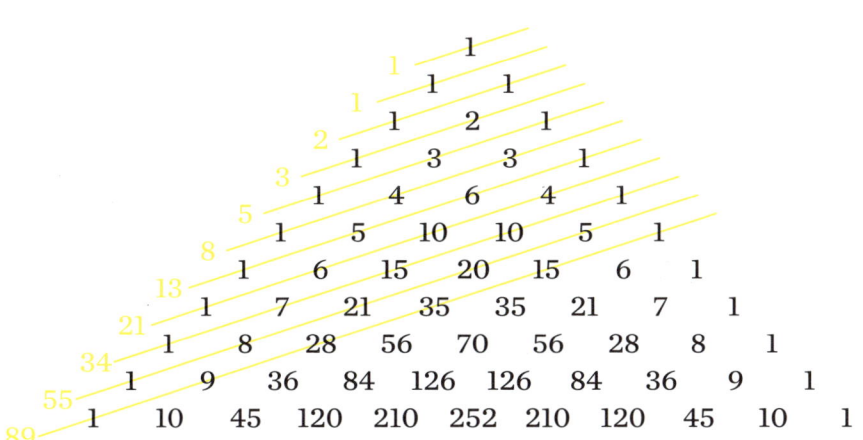

Here's one more beautiful property of the Fibonacci Numbers. If a, b, c and d are 4 consecutive Fibonacci Numbers, then $c^2 - b^2 = ad$.

Now it's time for you to use your little grey cells.

Go ahead and check that this is correct for each of the stretches of 4 consecutive Fibonacci Numbers that include this problem's number 13.

The slater (or woodlouse) has 14 legs

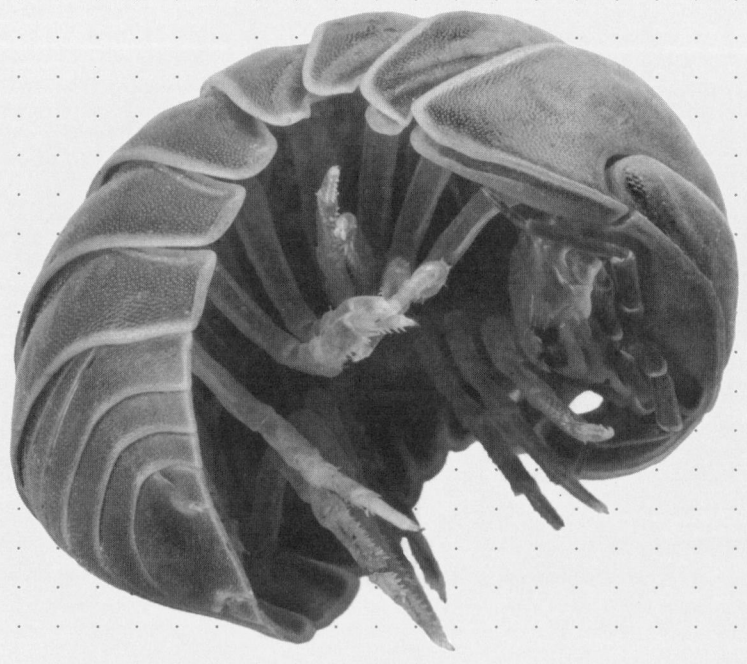

*But Adam, you say, all insects have 6 legs.
Correct! But these little critters are, in fact,
isopod crustaceans. They're actually more
closely related to prawns or lobsters than
insects and, yes, you can cook and eat them.
Apparently they taste like really, really,
really small shrimp.*

A Gematria Gem

The gentle art of substituting numbers with letters — such as A = 1, B = 2, C = 3 ... and so on — is variously called 'gematria' or 'numerology'. A nerdier name is 'alphanumeric substitution'.

Like any good pseudoscience, it allows for huge variation in interpretation, since it's entirely up to you whether, say, you include vowels or only focus on consonants ... not to mention which language you use.

Above is a portrait of the great composer Johann Sebastian Bach, who was reputed to dabble in a wee bit of gematria himself. He may have had a special interest in the number 14, which some suggest he 'encoded' in his music as a sort of signature. There are 14 canons included in Bach's appendix to the *Goldberg Variations*, for example, and he was rather fond of using pairs of 7s in his work.

But why might Bach be keen on 14? I reckon there are enough clues for you to strap your thinking cap on and see if you can nut it out.

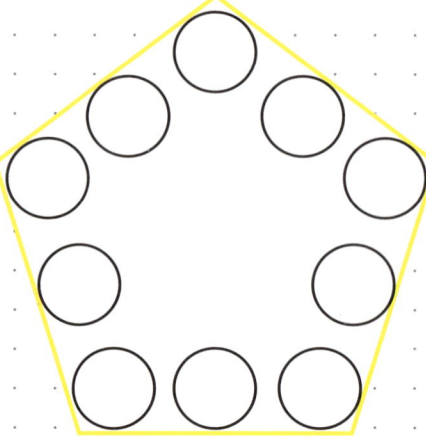

Pentagoing going gon

This pentagon has 10 circles on it: 5 at the corners and 5 in the middles of the edges.

Place the numbers 1 to 10 in the circles so that each side adds up to 14.

As this is the first of these you've seen, I'll give you a massive hint. Actually, I'll do most of it for you. Why are you being so kind, Adam? Trust me, you won't be asking that after the next few pages where these babies will be back!

The numbers 1 to 10 have to appear in all the circles and each side must add to 14. So if we add up the 5 sides we will get 5 × 14 = 70. But if you look at the diagram, 'adding' the sides will involve every circle once and the corners a second time. The circles contain the numbers 1 to 10 so adding the 10 circles up once gives us 1 + 2 + 3 + ... + 9 + 10 = 55. We need to get to 70 so when we add the circles a second time they must give us the extra 15.

How can 5 circles, each containing a different number, add up to 15? They must contain numbers 1, 2, 3, 4 and 5 in some order. That means 10 is in an edge circle, and the other two circles on the line through 10 must add to 4. These can only be 1 and 3, so one of the sides is 1-10-3. Similarly, 6 is in an edge circle, and the other two circles on the line through 6 must add to 8. These can only be 3 and 5, so one of the sides is 3-6-5.

See if you can fill out the rest from here using the same technique. There is only one possible answer — any others are rotations of reflections of it.

So go for it! Complete the pentagon.

Lazy circles

Put down your pencil and take a look at the two lots of circles above.
The top circle is 14 millimetres across. Without measuring (that is, just by looking — no fingers) see if you can estimate the diameter of the bottom black circle.

No cheating!

There are 15 species of Galapagos Giant Tortoise ...

... 11 of which survive to this day. But recently scientists have charted the DNA of several tortoises on Wolf Island near Ecuador and think they may have found some pure breeds or extremely close to pure breeds of a species thought to have been extinct, namely the Floreana Tortoise. It's hoped that a targeted breeding program could bring back this extinct species, but like anything a Galapagos Tortoise does, it will happen SLOWLY!

Oh and one other thing. While most people believe it was the finches on the Galapagos Islands that got Charles Darwin thinking about evolution, it was actually the various types of mockingbirds that he saw that made him realise that over time species can evolve, with a theory that is now accepted as part of the scientific canon in most places (cue banjos and letters of complaints!).

Tip-ical, just typical!

Basil and Bonita Beefeater like nothing more than belting down a $12.50 beef burger each over lunch.

Today they've done exactly that. They place $15 each on the counter and say to the waiter, 'That was great service , please take a tip'.

The waiter gives the $30 to the cashier to pay for the $25 worth of burgers.

The cashier hands back $5 in one dollar coins to the waiter.

The waiter keeps $3 as a tip and hands back $1 each to our satisfied Beefeaters.

So Basil and Bonita paid $14 each for the meal, which comes to a total of $28. The waiter has her $3 tip. But $28 + $3 makes $31. Where did the extra dollar come from?

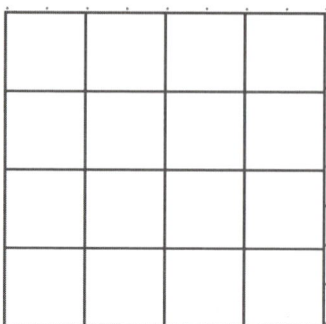

Another square affair

This puzzle is an old favourite that has tripped up thousands of students over the years because the answer seems so obvious.

How many squares are in the diagram above?

Most people answer 16 and move on. But that is not the case. Yes there are 16 small squares in the grid, call them 1 × 1 squares. But the grid also contains 2 × 2 squares, 3 × 3 squares and a big 4 × 4 square as its perimeter. I've highlighted some of those squares here for you.

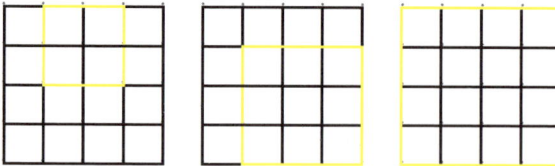

If you answered that there are 16 squares in the diagram, go back now and add in the 2 × 2 and 3 × 3 squares and the big daddy 4 × 4 and tell me now how many squares there are.

Can you see a pattern in how many 1 × 1, 2 × 2, 3 × 3 and 4 × 4 squares there are? If so, can you quickly work out how many squares are in the 10 × 10 grid on the opposite page?

Sixteen men died building the Sydney Harbour Bridge

I don't want to underplay the tragedy of 16 workers dying, but given the immense scope of the project and the incredibly dangerous conditions in which they worked, it seems a stunningly small number.

Incredibly, Vincent Kelly, who fell over 50 metres and hit the water with such force as to send foam 7 metres into the air, SURVIVED. After several somersaults he managed to hit the water feet first, then swam 50 metres to the punt in the harbour which ferried him to safety.

Takes swan to know one

During the year, I got this email from a sixth-grader name Yexuan.

Subject: *Maths*

Message: *This has nothing to do with your book, Adam, but I was wondering if you could answer this question for me: In the well-known ballet 'Swan Lake', there are usually 16 'swans'. But in a new production, there are 48 swans.*

What percentage increase is that from the usual number?

Half of our class says it is 300% and the other says 200%. Please give me an answer before my class starts a maths war!

If you had been in Yexuan's class and the maths war broke out, on which side would you be fighting?

Check to see if you'd be on the winning side ... at the back of the book.

Seventeenth-century quake-buttocks

*The 17th century was a cracking one for words that have fallen out of use, but I think should be brought back. Take for example the great word of the time to describe a coward. Next time someone is
a bit short of courage, why not call them a quake-buttock?*

Other Awesome But Very Obscure 17th Century Words I'd Love You To Help Me Get Back Into Circulation include (but are not limited to ...)

Cunctation: the act of avoiding something or putting it off.

Jargogle: to confuse things or jumble them up.

Twitter-Light: another word for twilight. Twitter-light was last popular in the 17th century but should be brought back because it's certainly around sunset that my daughters start hitting the social media. Maybe Insta-light or Snap-light would be more fitting?

Widdendream: an old Scottish way of saying you were mentally disturbed or confused was to say that you were 'in widdendream'.

Who's with me? Come on! Don't be a quake-buttock! Now is not the time for cunctation. Snap out of your widdendream and help bring these classics back.

(Fermat) prime time

The 17th-century philosopher, scientist and moustachieoed maths man Pierre de Fermat was a total arithmetical gun. He had many ideas that changed the world of mathematics — and even gave a great explanation as to why we see rainbows. But not every idea he had was correct.

One Fermatian foul-up occurred when Pierre considered numbers of the form $F_n = 2^{2n} + 1$, which we call Fermat numbers.

Now don't be freaked out by the look of this. When $n = 2$ we have that $2^2 = 4$. So the Fermat number F_2 is given by $F_2 = 2^4 + 1 = 16 + 1 = 17$.

Now, $F_2 = 17$ is prime, as is $F_1 = 2^2 + 1 = 5$ and even $F_0 = 2^1 + 1 = 3$. The example of F_0 only makes sense if you remember that $2^0 = 1$ from high school. And if that really confuses you, it might be best just to let it slide for now.

The point is F_0, F_1 and F_2 are all primes. So are F_3 and F_4. This was enough for old Pierre to get very excited and pronounce, in 1650, that 'all Fermat numbers are prime'. Oh Pierre, Pierre, Pierre!

The greatest mathematician of all time, Leonhard Euler, rolled up his sleeves and showed that in fact $F_5 = 4294967297 = 641 \times 6700417$.

In the age of just quills and ink, it was tough to expect anyone to do the larger calculations, but it turns out that:

$F_6 = 18446744073709551617$, $F_7 = 340282366920938463463374607431768211457$, and F_8 ... wait for it ... $F_8 = 115792089237316195423570985008687907853269984665640564039457584007913129639937$ are also not prime. They can all be factorised (again file this under 'best just let it slide for now').

Okay, okay, since you insist ... in 1855 (again, without a calculator), Thomas Clausen showed that $18446744073709551617 = 274177 \times 67280421310721$.

And it was only in 1970 and 1980 that we realised that $F_7 = 59\,649\,589\,127\,497\,217 \times 5\,704\,689\,200\,685\,129\,054\,721$ and that F_8 can be factorised as the jaw smashingly big $1238926361552\,897 \times 93461639715357977769\,163558\,19960689658405123754163818858\,0280321$.

We've actually gone as high as F_{32} which I get lightheaded even thinking about, and shown that none of them above F_5 are prime. So, Fermat, sorry buddy, but you got that one wrong!

Why have I brought you to this migraine-inducing point? Well, I just want you to go back and find the biggest two Fermat prime numbers we do know, F_3 and F_4 ... preferably by hand!

Snake's alive!

On 24 May 2017, Nokia re-released its classic 3310 phone, 17 years after it originally went on sale.

By all accounts it jumped off the shelves and Nokia couldn't keep up with demand!

The new edition had massively better screen definition, battery life and a 2MP camera, but most importantly still had the original phone game from way back before Angry Birds or Pokémon GO ... I'm talking 'Snake', baby!

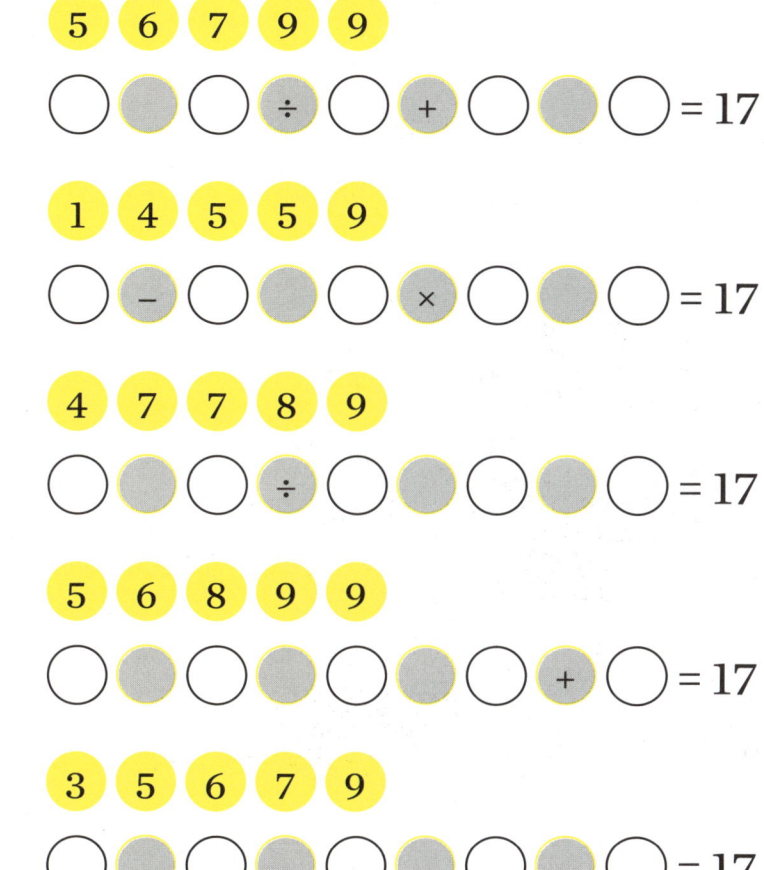

5 6 7 9 9

◯ ⬤ ◯ ÷ ◯ + ◯ ⬤ ◯ = 17

1 4 5 5 9

◯ − ◯ ⬤ ◯ × ◯ ⬤ ◯ = 17

4 7 7 8 9

◯ ⬤ ◯ ÷ ◯ ⬤ ◯ ⬤ ◯ = 17

5 6 8 9 9

◯ ⬤ ◯ ⬤ ◯ ⬤ ◯ + ◯ = 17

3 5 6 7 9

◯ ⬤ ◯ ⬤ ◯ ⬤ ◯ ⬤ ◯ = 17

Operator, operator!

Here's a fun type of puzzle invented by fellow numbers nerd Sean Gardiner. We'll visit these twice more in our number games, so let's start with a fairly simple one. When I say 'simple', what I really mean is 'the easiest of the three in the book'.

So limber up and see if you can figure out where the numbers fit, and which operators they require, to complete the equations above. Too easy? Then see if you can find the unique arrangement that gives the highest 'score' in each line, where your score is the 5-digit number you get from reading off the yellow numbers in order!

Qaanaaq

In mid-2017 the famous Australian polar explorer Geoff Wilson and his son-in-law Simon Goodburn walked the length of Greenland, unsupported, from south to north, in only 18 days.

To put this in perspective, the old record was 42 days. In temperatures as cold as minus 20 degrees Celsius, they used kites to drag them and their 150-kilogram sleds across the 2160 kilometres to end up at the palindromically-named Qaanaaq, one of the world's northernmost civilian settlements.

And in the most Aussie part of all, they come from the Gold Coast and practised by hauling tyres across a beach during summer!

Never gonna give you up ...

This is my fourth book in the series (the other three are available at adamspencer.com.au and good bookstores everywhere ... PLUG!).

I thought I'd give a shout out to a classic. In my *Big Book of Numbers* I called it 'A Rickety Ol' Problem', so lets Rick-roll again with this hall-of-famer.

Of all the problems I've posed in my previous three books (*which are available at adamspencer.com.au and good bookstores everywhere!*) this one has received the most responses — from people who just can't get the answer, from people so excited that they did, even from my old radio buddy Wil Anderson who once worked it out on stage in the back of his mind while doing a 90-minute stand-up gig!

Amy, Ben, Cassius and Delia need to cross a rickety old wooden bridge between two mountain tops late at night. Don't ask why, that's not important. They only have one very weak torch which means only two of them can cross at any time. In addition, when two people cross together, they have to travel at the slowest person's pace so they can both see. And to top things off, in exactly 18 minutes the bridge will collapse (again, see 'don't ask why, that's not important').

Amy can get across the bridge in 1 minute, Ben 2 minutes, Cassius takes 5 minutes and Delia 10.

If Amy and Delia go across, that takes 10 minutes and Amy returns with the torch taking 1 minute.

If Amy and Cassius then cross and Amy returns, it takes 5 minutes plus 1 minute. Amy and Ben then cross together in 2 minutes.

In total, that takes 10 + 1 + 5 + 1 + 2 = 19 minutes and just about anyone you ask comes up with that as the quickest possible crossing time.

But in fact they can get across quicker. Not in 19 minutes ... not in 18 minutes, but *under* 18 minutes.

Find a combination of crossings that gets Amy, Ben, Cassius and Delia safely across the river in less than 18 minutes.

Trust me, there is an answer — it just involves going against your most basic impulses in trying to solve the question.

Oh, and consider yourself Rick-rolled.

Sydney FC, FTW

In an incredible 2016/17 A-League football season, Sydney FC broke multiple records including going unbeaten for their first 19 games.

Other records include 66 (most competition points in a season), 11 (most goals to start a season before conceding one), 17 (most points ahead of second place at season's end), 12 (least goals conceded in a season), 16 (most clean sheets in a season), 43 (more goals scored than the opposition), and 1 (number of games where they conceded more than one goal) ... what a year!

While you're here, see if you can spot the 'error' with the soccer ball pattern above. Focus on the black shapes ... can you see it?

Yes, my eagle-eyed, geometry-loving, football fanatics ... the black shapes on a soccer ball should be pentagons, not hexagons!

8, 18, 11, 15, 5, 4, 14, 9, 19, 1, 7, 17, 6, 16, ?, ?, ?, ?, ?

A logical jumble

I've written the numbers 1 to 19 in a specific order above.
It's probably obvious to you that the numbers I've still got to write down are:

2, 3, 10, 12 and 13

But in which order should I write them ... and why?

$$2 \times 2 \times 2 \times 2 \times 2 \times 2 \times 2 \times 2 \times 2 \times 2 \times 2 \times 2 \times 2 \times 2 \times 2 \times 2 \times 2 \times 2 \times 2 \dots phew!$$

If you took these 19 twos, multiplied them all together and subtracted one, you'd get the number 524287.

In 1588, Pietro Cataldi proved this was a prime number, simply by exhaustively checking by hand if it was divisible by every prime up to 719 (719 being the last prime before the square root of 524287, meaning any factor greater than 719 must multiply another number less than 719). Nice work, PC!

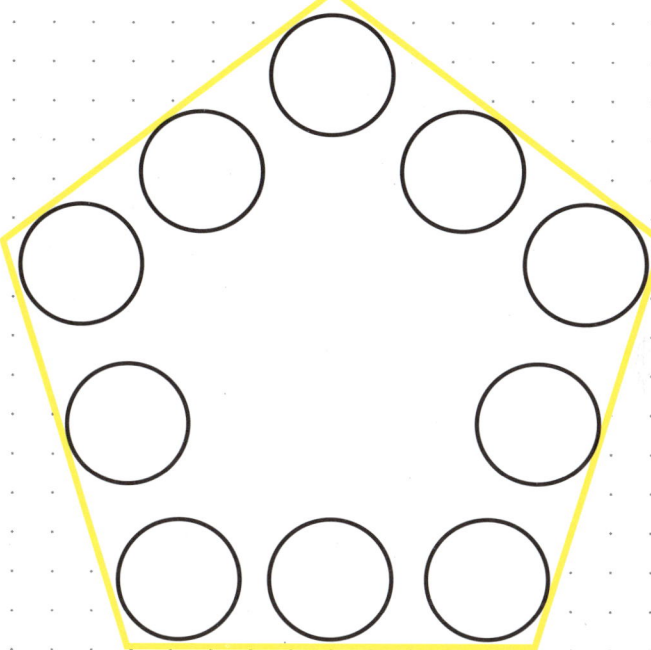

They're baaaack

H ere's another puzzling pentagon, similar to the one we gave you for the number 14.

This time I'd like the numbers 1 to 10 placed in the above circles so that each side totals exactly 19.

If you haven't done the puzzle for 14 yet, go back and try it. It shows you the logic to follow here.

Oh, and if you liked this puzzle, there is a much, *much* harder version waiting for you at 85.

If you've ever wondered ...

how many human sperm you could fit, head-to-tail, in a normal-sized full stop, in an average book, wonder no more ... it's about 20.

R	O	U	F	O
U	F	R	O	U
U	O	F	F	R
R	U	R	O	F

Tetrocharged

You might remember meeting the 12 free pentominoes a few pages back (and if you don't ... well, you know the drill ...)

If you only have 4 squares to play with (I know budget cuts suck, don't they!) you can form what are called 'tetrominoes'.

There are a mere 5 free tetrominoes ... find them.

Got 'em? Good!

Now, divide the grid above into 5 tetrominoes, so that each tetromino contains the letters F O U R in some order.

Find at least two solutions. BTW, it's impossible to tile a 4 × 5 grid with all 5 different free tetrominoes, so expect to see some repeated pieces.

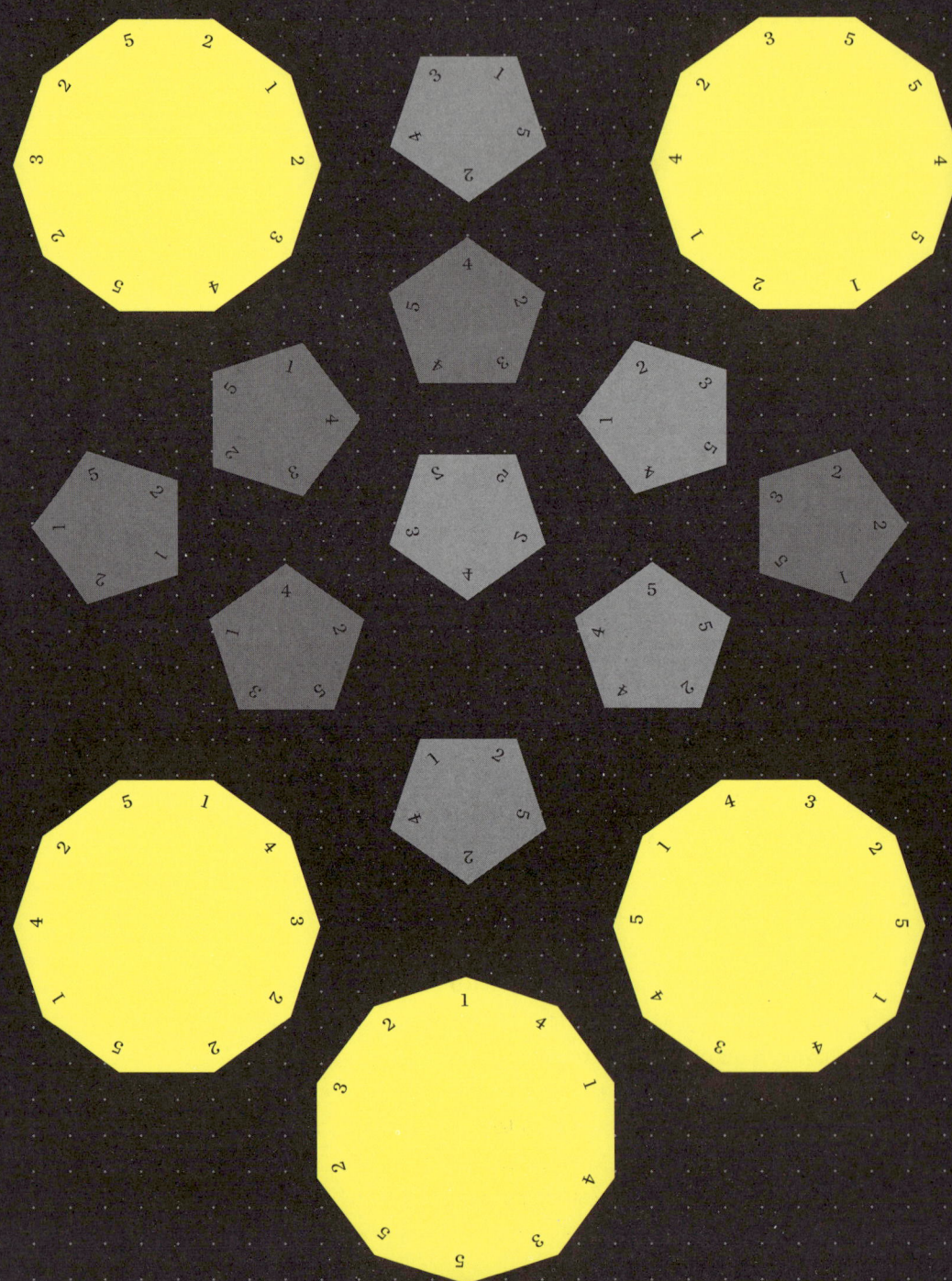

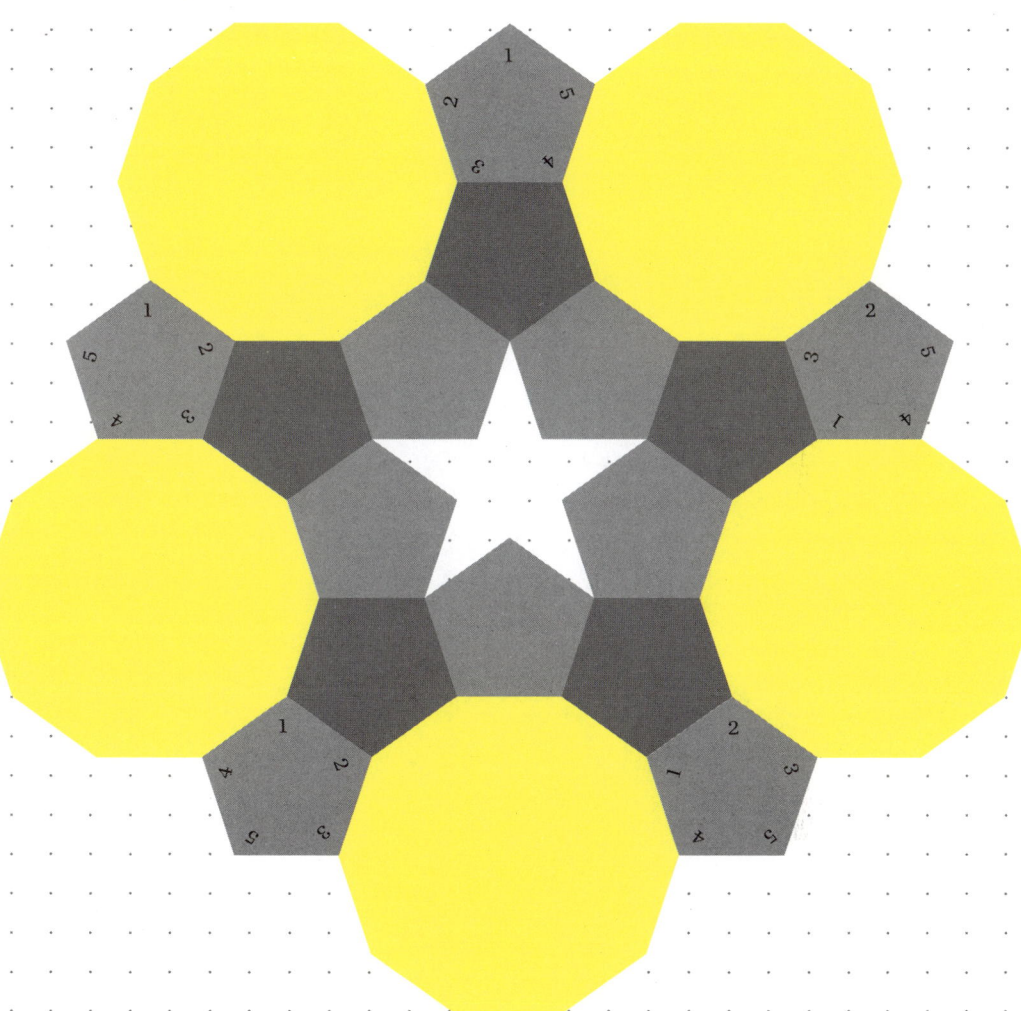

Aa ha!

Check out this gorgeous 20-tile tesselation!
Stay tuned until the next page to learn more about it, but in the meantime, why not see if you can fit the puzzle pieces from the opposite page in? Every common corner should match the same number.

You can download the shapes from my website for free if you'd prefer to keep your Tetrocharged answers for posterity. And hey, why wouldn't you?!

Head over to www.adamspencer.com.au and grab your scissors and gluestick!

020

Say 'Aa'

The grid we used on the previous page for that fiendish puzzle was discovered by that mathematical marvel Johannes Kepler.

Yep, he's the dude who worked out that the planets orbit the Sun, not in circles but by sweeping out ellipses. He also figured out that the most efficient way to stack a pile of cannonballs (without knowing which, you may be confused for our heading to 29!).

Kepler discovered this pattern while investigating whether you could tile the plane with shapes including pentagons. He came up with the beautiful design which he simply labelled 'Aa' in his cracking read Harmonices Mundi.
If we extend the diagram beyond the puzzle I set you, you can see that it also involves 5-pointed stars called pentagrams, as well as decagons and the curious 'fused decagon' which you can think of as two decagons getting a little bit friendly with each other.

Kepler clearly rated his discovery, writing that the 'structure is very elaborate and intricate'.

021

Teeth were once a common birthday present

Well into the 1900s, especially in areas affected by poverty, a common 21st birthday present was for someone to pay for you to have all your teeth taken out and replaced with dentures. This would save on dental costs throughout your life. The dentures may even have contained real teeth, taken from the mouths of soldiers who had died in battle.

A Poe-posed cipher

The following cipher was sent to the famous author and puzzler Edgar Allan Poe by 17-year-old Schulyer Colfax who went on to be equally famous in different circles.

In fact, Schulyer was the 17th vice-president of the United States. He opposed slavery (pretty cool at the time) and once toyed with joining the 'Know Nothing Party'. It's a weird name for a political party, but it used to meet secretly and when people were asked if they were members, they were expected to answer 'I know nothing'. The name stuck!

Anyway, this code looks absolutely awful at first, but see if you can translate the 21 different letters, numbers and symbols into letters to give a phrase.

There are no unusual words in the answer, but given that the puzzle was set in 1840 I should warn you that the language is fairly 'formal'.

As another hint, two of the letters stand for themselves and the semi-colon ';' and the full stop are actually pieces of punctuation. Are you ready? Like I said, it looks hideous!

8n()58†d w!0 b† !x6n†z k65 !nz k65,8l†n b)x 8nd)Pxd !zw8x 6k n6 36w-†nd!x86n;

x=†0 z†,5!z† x=† w8nz 8n 8xd 62n †dx††w !nz k653† 8x x6 5†36l†5 8xd P†l†P b0 5†l†n,†.

()n8)d

This is a pretty tough question. If you're tearing your hair out, a logical approach will help you bust this open. Firstly, go through and make a list of each symbol (letter, number or punctuation mark) and write down how many times they occur. Then think about which letters occur most commonly in English (or look up a list if you want). There is one symbol which clearly occurs the most often and a 6-letter word in the cipher in which this symbol occurs 3 times including in a double. What word could that possibly be? Go from there.

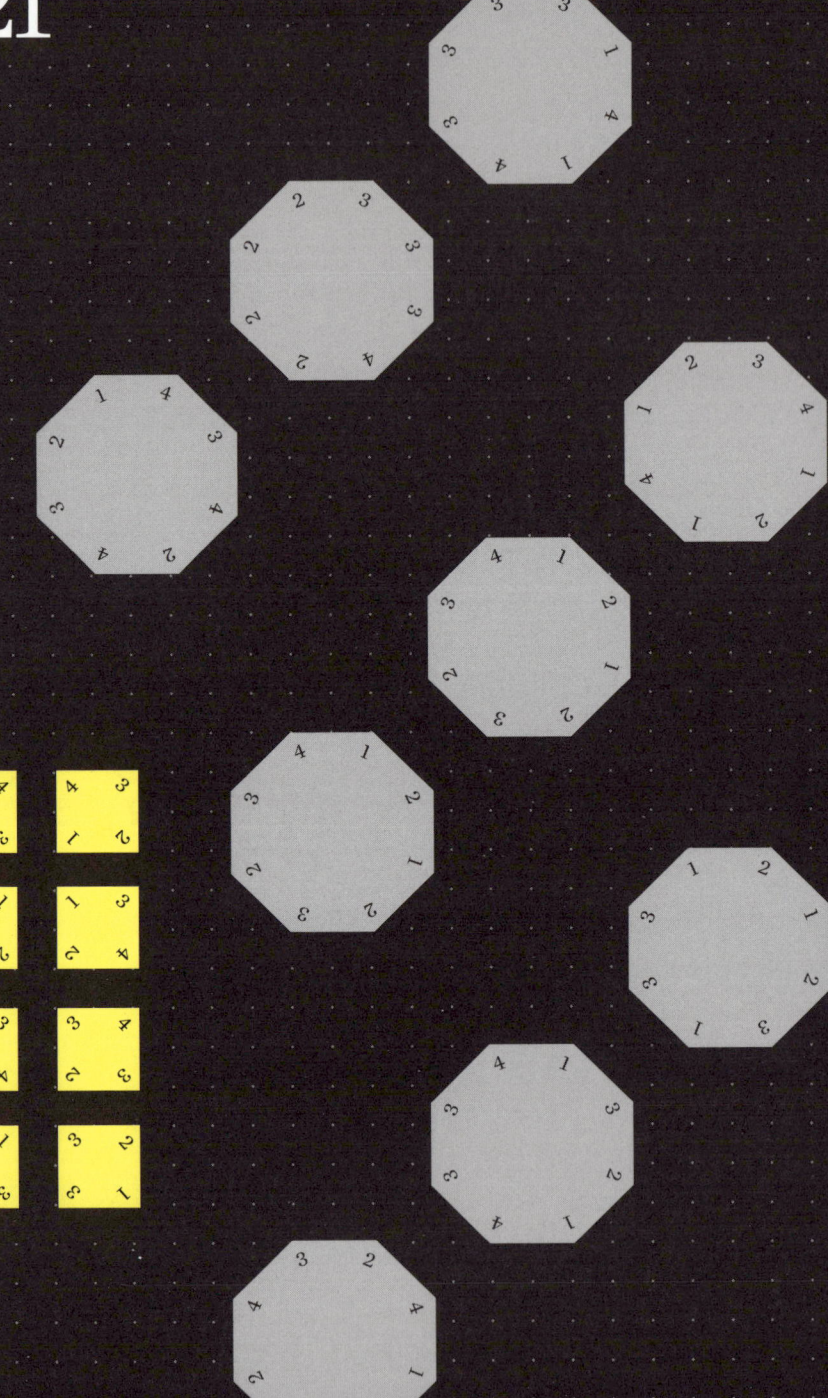

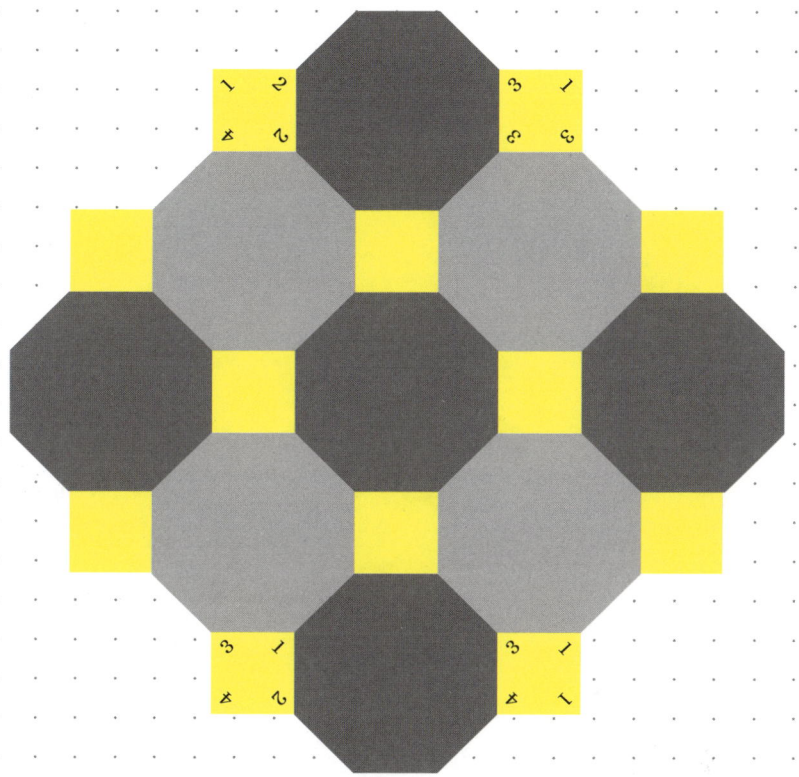

Easy-peasy?

Who doesn't love a good tesselation?

Well, you can admire this 21-tile beauty as you figure out how to place the squares and octagons from the opposite page into the pattern above, so that every common corner shows the same number.

Before you go and grab the scissors and attack the book (which, of course, you're more than welcome to do ... so long as you own it!) you can download and print out the shapes for free from my website.

But my generosity doesn't end there. I've given you 4 of the squares to start you off. 'Gee, thanks Adam ...' No worries!

What do you call a snake that's almost exactly 22/7 metres long?

A πthon.

Partition party!

In mathematics when we 'partition' a number, we break it up into smaller chunks that add together to give us that number.

So the partitions of 5 are:

5, 4 + 1, 3 + 2, 3 + 1 + 1, 2 + 2 + 1, 2 + 1 + 1 + 1, 1 + 1 + 1 + 1 + 1.

Note that we usually write partitions from largest to smallest and 4 + 1 is the same as 1 + 4 so we don't double up on that sum.

The 'strict' partitions of 5 are 5, 4 + 1 and 3 + 2, because they don't repeat any numbers in the sum. So there are 7 partitions and 3 strict partitions of the number 5.

As numbers increase, the number of partitions of them increase too — and quickly. The number 43, for example, has 1610 strict partitions.

You might be thinking you can feel a question coming on — and you'd be right.

Why don't you find the number less than 10 that has 22 partitions?

And find the number in the mid-teens that has 22 strict partitions.

Party on!

On 22 April 2017 ...

The awesome Cassini Spacecraft started the first of its 22 dives through the rings of Saturn in what NASA describes as The Grand Finale.

Cassini was launched in 1997 and has been bringing us close-ups of Saturn since 2004. The Finale will come with Cassini eventually plunging into Saturn's atmosphere on 15 September 2017. What a way to go!

In the first verse of the Jay-Z song 22 Two's ...

He uses the word 'too', 'to' or 'two'... you guessed it ... 22 times.

Unfortunately the song is not in any way mathematical, and suffice to say the lyrics are far too profane to reproduce in this family-friendly book.

22

A 23-year-old woman who wanted financial support from her parents was told to take a good long hard look at herself by a Spanish judge

In 2017, in a landmark legal ruling, a 23-year-old Spanish high school dropout clarified the law in Spain where it is expected that parents will provide for their children until they have become financially independent. The law puts no number on this and it's led to the phenomenon of so-called 'parasite children'.

This parasitic progeny was put particularly in her place in a scathing judgment that said she was lazy, neglectful and was 'wasting her life'.

No prizes for guessing who took the garbage out and did the washing up that night!

1	5	3	0	4	6
1	4	0	2	3	5
1	2	2	0	1	5
4	5	5	2	1	7
3	2	2	3	5	9
9	9	3	1	9	14

3	5	4	3	3	3
4	4	3	5	5	4
4	3	5	4	3	5
4	3	4	4	4	3
4	5	3	4	4	3
4	3	4	3	4	5

Divide and concur

For each of these grids draw 3 straight vertical and/or horizontal lines that divide the grid into 6 sections, so that the numbers in each section add up to 23. Answers at the back of the book, of course!

Smile!

If you've ever had your photo taken with a Gatso 24, chances are you weren't actually smiling. The GATSO Type 24+AUS Speed Device is one of the world's most used speed cameras.

Non-transitive Dice

Dice with 6 sides, labelled one to six are so yesterday, man. Kira, Bishan and Carlitto have 3 nifty dice, labelled as follows:

Akira's Die A has sides 1, 1, 3, 3, 8, 8
Bishan's Die B has sides 0, 0, 5, 5, 7, 7
Carlitto's Die C has sides 2, 2, 4, 4, 6, 6

The faces of each die total exactly 24, but no two dice contain a common number.

They play a game where two of them face off and roll their dice. The owner of whichever dice shows the highest number wins a point and the winner then rolls against the other player who didn't roll that round. This goes on and on for a long time. A REALLY long time.

Who would you expect to win after, say, 1000 rolls?

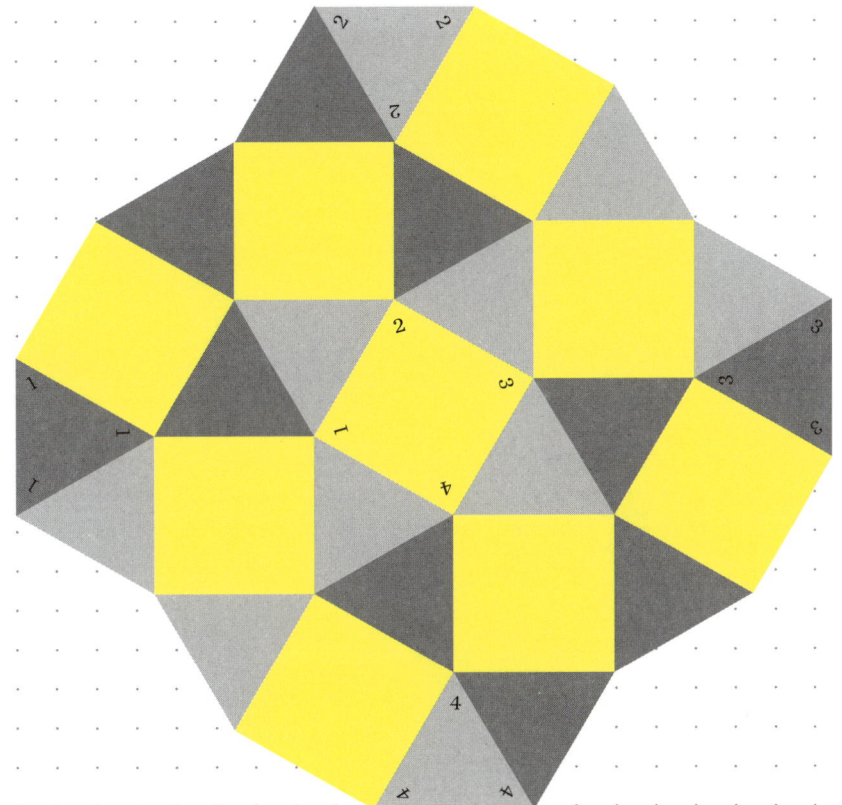

Mile of tiles

Here's one to bear in mind next time you renovate your bathroom. Hey, why not? I reckon this gorgeous tesselation would look great on your bathroom floor. But I digress.

The idea is to place the 24 tiles on the left onto the template above. Each number must match at every common corner (1s with 1s, 2s with 2s, and so on). As with many of these sorts of puzzles, I've got a downloadable version on my website if you'd like to print it out and cut it out.

Get tiling!

A number of words ...

Here you'll find three 25-square grids. Each contains 5 columns of five letters.

Taking one letter per column as you move left to right across this first grid, spell five 5-letter words. Here's a hint: each of the words is a number.

T	E	X	E	T
S	I	R	E	N
E	H	V	T	Y
F	I	R	T	E
S	O	G	H	Y

This next grid, on the right, contains another five 5-letter words from a common group. They are the five largest 5-letter things of this group in the world, and as a further hint ... everyone in the world lives in one of these types of things.

C	A	Y	A	T
E	N	I	P	A
I	H	A	I	N
J	G	D	N	Y
I	T	P	L	A

This final grid contains five incredibly well-known 5-letter things, all from the same category ... but you're not getting any more hints!

P	P	P	P	O
M	E	A	O	E
G	E	N	L	N
L	R	M	C	E
A	A	A	G	H

25
YEARS

According to some ...

including a table on Wikipedia (so it must be true!), the 'preferred term' for a 25-year anniversary is a 'quadranscentennial'.

I'll be honest: I've never heard it used outside of this list. The other term offered — silver jubilee — is much more common.

But if you're a jargon collector, go right ahead and stick this one up your sleeve to wow your mates. Whatever you do, don't call it a quadricentennial — which is a 400-year anniversary!

Pretty dicey

You might still be reeling (or rolling) from meeting the 'non-transitive dice' listed as our puzzle for the number 24. Well, hold on to your temples!

Let's say we have 5 dice (A, B, C, D, E), labelled as follows:

A:	R	2	2	7	7	7
B:	R	1	6	6	6	6
C:	R	5	5	5	5	5
D:	R	4	4	4	4	9
E:	R	3	3	3	8	8

For each die the numbered sides total 25, and if you roll an R — that's a 're-roll' — you simply roll again.

You might remember from the three dice we saw at 24 that we could make a chain where A beats B beats C beats A (by 'beats' here I really mean 'is more likely to win against'). Looking at the five dice above, you should be convinced that A beats B and B beats C, but C doesn't beat A here. Can you complete the chain 'A beats B beats C beats...' to get back to A? Are there any other chains like this?

J, the 26th letter?

For a long time English had 25 letters, until J came along as the last letter to join the alphabet sometime in the 1600s. So in a bizarre way, J is the 26th letter of the alphabet! Here are some cool facts about the 10th (26th?) letter ... J.

I love the letter J. So much that my younger daughter Olivia has the middle name JJ. She actually renamed herself Olivia Diamond JJ Spencer, but that's another story.

'J' started life as an 'I' and was used in Roman numerals as the last 'i' in a number containing multiple 'i's. So, 23 was sometimes written xxiij not xxiii. It first started to evolve as a letter in the works of Italian Gian Giorgio Trissino in the 16th century.

In Scrabble there is only one letter J, and it's worth an impressive 8 points (the same as X, but not as much as Q and Z, both worth 10). Note that if you're playing Scrabble in Dutch you'll only get 4 points, in Czech 2 points and in Hawaiian ... you'll get no points since Hawaiian ain't got no J in it.

The only Scrabble-certified 2-letter words containing J are Ja, a version of 'yes' from German, and Jo, being a Scottish term for a sweetheart, and there are a mere three 3-letter Scrabble-certified words that end in J, namely Haj, Raj and Taj.

J used to be the only letter not to appear in the periodic table of chemical elements. It now shares this distinction with the letter Q, which was wiped from the table in 2012 when the placeholder element name ununquadium (Uuq) was officially updated to flerovium (Fl). J had a fighting chance to appear on the table in 2016, when element number 113 was announced to be named after the country of Japan... but instead of 'japanium' it became 'nihonium', based on the country's native name Nihon. Better luck next time!

The dot above a j or an i is called a tittle or 'superscript dot', and the curve away at the bottom is called a 'tail'.

According to an analysis of words that appear in the Concise Oxford English Dictionary, J is the second least used letter in the English language. The full list, in order ... EARIOTNSLCUDPM-HGBFYWKVXZJQ. Other lists have J pushing up past X and Z into 4th last place. Regardless, she's a rare one.

The radio station JJJ or Triple J is without doubt the best radio station in the world.

The gambler's fallacy

The 'Gambler's Fallacy', also sometimes known as the 'Monte Carlo Fallacy', is a common mistake people make when gambling on things like the toss of a coin or the roll of a roulette wheel.

The mistake is to see one result happening many times in a row — like a string of heads or red numbers on a wheel — and think that the chances of a tail or black spin must be going up because 'this run just can't keep going'.

This is an easy mistake to make when caught up in the excitement of watching someone toss 5 heads in a row but if you think about it, when I toss a coin or roll a ball into a roulette wheel, the coin or the ball doesn't 'know' what the last result was, and it certainly can't choose the way it comes up next. If I've tossed 9 heads in a row the odds for tails on the next toss are still 50-50.

Probably the most famous example of the Monte Carlo Fallacy occurred on 18 August 1913 at the Monte Carlo Casino in Monaco. According to the legend, the roulette wheel spun up a black number a whopping 26 times in a row and gamblers lost millions of francs betting big on red thinking 'it must come this time!' Years later, in 1943, an American casino recorded 32 straight red spins.

So, what would you guess are the odds of tossing 26 heads in a row with a fair coin? I don't expect an exact answer, but have a go. Would it help if I told you that the odds of 10 straight heads is 1 in 1024?

The longest appendix ever removed was from a Croatian patient in Zagreb and measured 26 centimetres

The average length is more like 9 centimetres.

The most amazing appendix ever removed came from the lower abdomen of Russian doctor Leonid Rogozov. What makes it so incredible was he removed it himself! He was in the Antarctic in 1961 and when he got acute appendicitis he knew there was no way he could get to a hospital anywhere else on Earth before it would burst and most likely kill him.

So L-Rog applied a local anaesthetic, got two colleagues to hold mirrors and pass him the instruments, and snip snip , away he went. Two hours later he had completed one of the gutsiest (or appendixiest) feats in the history of surgery.

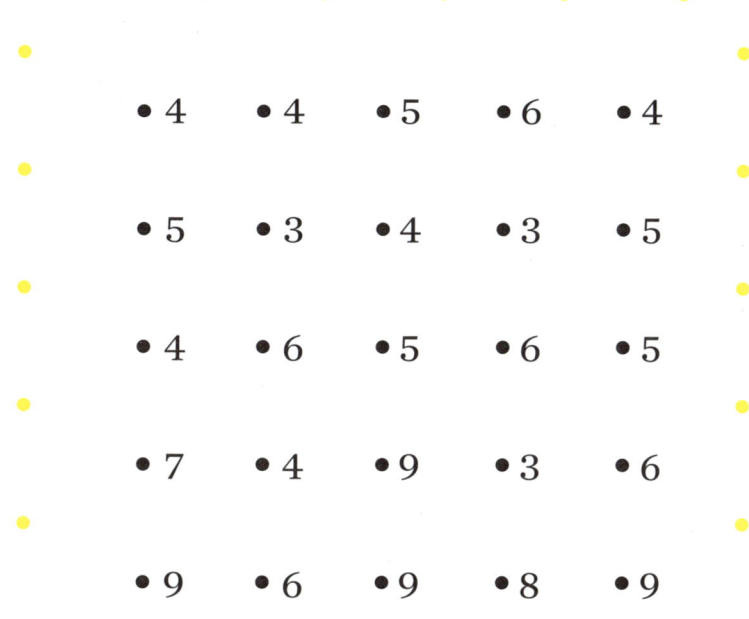

Splice, splice, baby

Split the array of numbers above into sections so that each non-empty region's numbers sum to — you guessed it — 26.

But wait! There are a few rules before you get started:

- To mark off these regions you may only use 3 lines.
- Each line must be straight and run between 2 yellow dots.
- If a line goes through a black dot, that dot's number has been crossed out and doesn't belong to any region.

Note it's the dots that score the points. If a dot is in a region it keeps its number even if the number is crossed out or lies in another region. Get it? Have fun!

Twenty-seven millimetres, thanks ...

When Australia's national football team the Socceroos play at the Homebush Stadium in Sydney, they request the grass be cut to 27 millimetres in length, according to head groundsman Graeme Logan.

27mm

Cubist painting

I have a cube whose edges are each 30 centimetres.

I paint the entire outside of the cube yellow. Then I cut the cube into 27 smaller cubes each of side length 10 centimetres.

These smaller cubes will have some yellow on them.

How many of the smaller cubes will have an odd number of yellow faces?

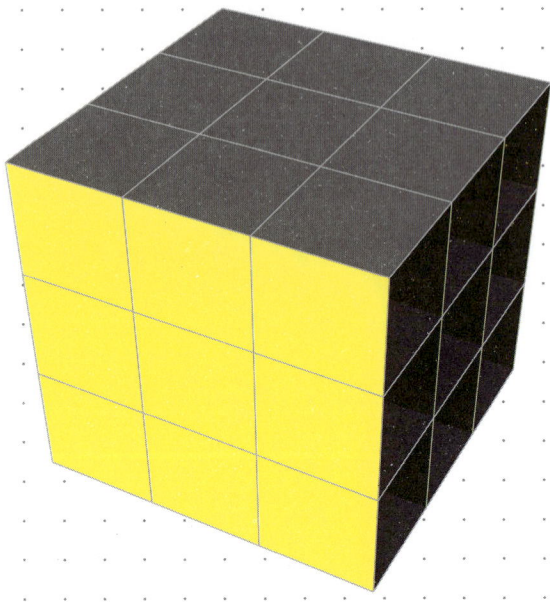

Menger management

Start with a cube made up of 27 smaller cubes like you see above. Now remove the middle square on each face and 'burrow through' to the other side — giving us a hollowed-out cube like the following.

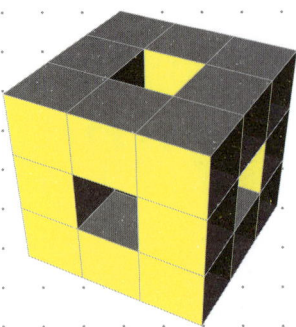

How many of the original cubes remain?

Keep going and remove the centre square of each of the 8 squares on each remaining face and burrow through, like this:

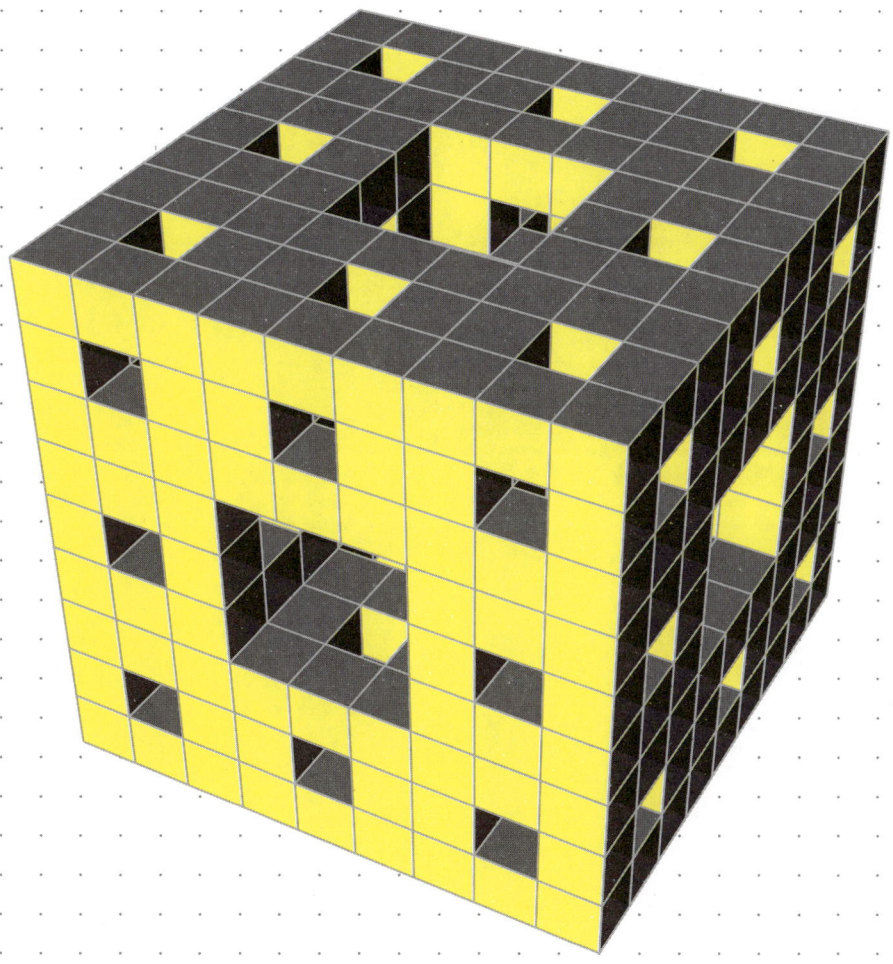

And so on ...

If you keep doing this FOREVER you get what is called a Menger Sponge.

Don't try and draw one, but picture one in your mind.

Ask yourself this: what is the volume of a Menger Sponge? And what is its surface area?

London's Natural History Museum houses a collection of over 28 million insect specimens

But in 2008 they discovered a new one ... outside. In their own garden to be precise. No less than the museum collections manager's 5-year-old son, while having a picnic in the museum gardens, found a small red and black bug about the size of a grain of rice that didn't match anything in the museum's records.

Grid 1

5	+	6	×	4
+	5	×	5	×
8	+	8	+	7
×	7	−	8	=
8	×	9	=	28

Grid 3

5	+	8	+	2
×	5	×	6	+
6	×	2	−	7
×	7	+	6	=
6	×	4	=	28

Grid 2

9	×	5	×	7
−	9	×	5	×
8	+	2	×	3
−	8	−	4	=
6	×	5	=	28

Number snake

Remember Snake from 017? Well, here's our far mathsier — and cerebral — version for you.

In the grids above, make a path from the top left corner to the bottom right by moving between adjacent squares. You can move up, down, left, or right, but you can never visit the same square twice. And, you guessed it, the path has to trace out the correct equation to get the final answer — 28.

There are 3 answers for each of the grids. I'll even give you a hint for the first grid. In each of its three answers, the path *always* starts by going down. Have a go at finding all 3 solutions to each grid. Even if you can only find one path for each grid, I'll be super impressed. Order of operations applies (so 3 + 4 × 5 = 3 + 20 = 23)!

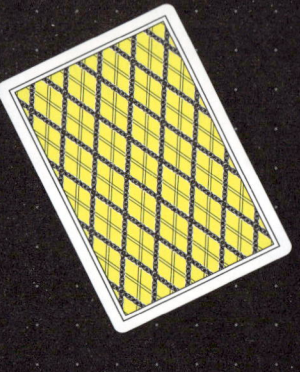

If you're ever hanging at your grandmother's

… while she and a few friends bang out a quick game of cribbage, and you hear some serious screaming, it could be because an incredibly rare 29 hand has been played.

This hand features four 5s and the Jack of Nobs; nobs being the suit that has been turned up in the hand.

When I say 'incredibly rare', I mean it — a 1 in 216580 chance - that's approximately 0.00046%. Not surprisingly, most cribbage players never see one.

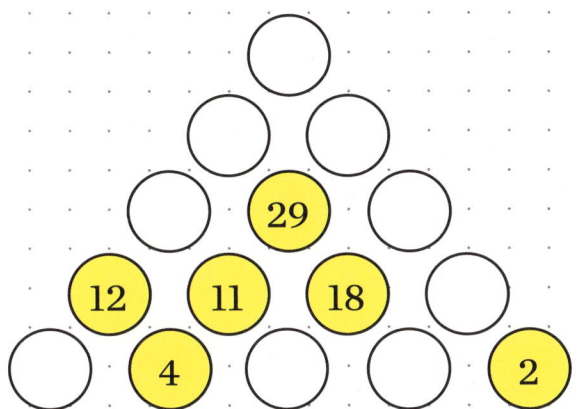

Eat yer heart out, Kepler

Waaay back in the 17th century, the great Johannes Kepler devised an awesome theorem about packing spheres in a three-dimensional Euclidean space.

Think of cannonballs arranged in a neat pyramid, and you're on your way to visualising what that means. But if you'd care to read up on it a little more, I'd suggest you check out the AMAZING book *World of Numbers* by, ahem, yours truly. But I digress!

In the number pyramid above, each circle is filled by the number that is the sum of the two numbers directly below it.

So 11 + 18 gives us the 29 in the circle above the 11 and 18. What number goes in the top circle?

The world's geekiest address?

Some people might suggest the nerdiest address in the world is the block of apartments 2311 North Los Robles Avenue, Pasadena, California.

You know it?

Next to a store that sells lamps? About a block from Pasadena City Hall? At least 5 storeys high and containing 16 or more apartments from what we can tell when we see shots of the mailboxes? Leonard, Sheldon and Penny live on the 4th floor? Yes, I'm talking about the apartment block in Big Bang Theory.

But even this nerd factory on the corner of Woodbury Road has nothing on The Cavendish Laboratory, AKA the University of Cambridge's Physics Department. In total, not one, not two, not three ... but 29 Nobel Prize Winners have worked there!

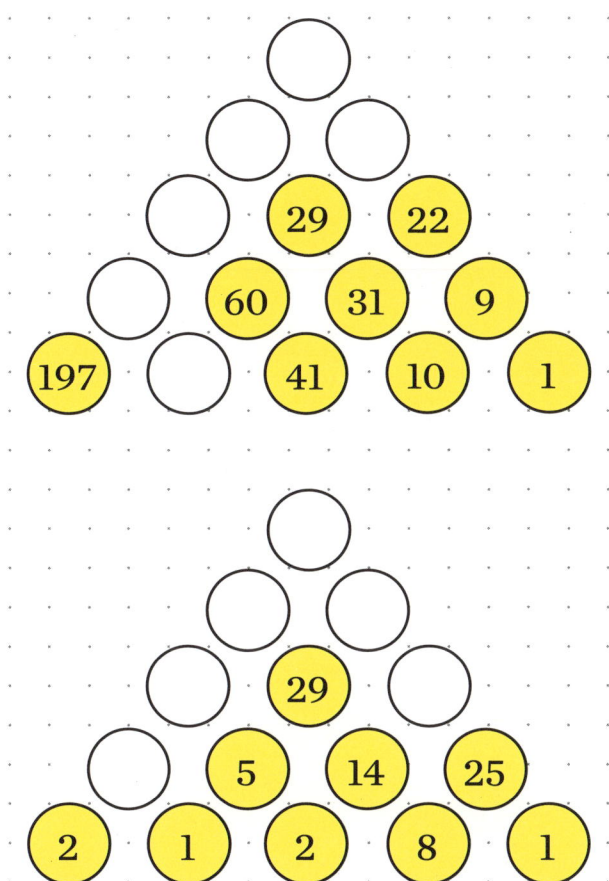

Stacks more

Okay, so now you know the drill, have a go at these ones. In these pyramids, again the value in any circle is determined by the value in the two circles below it, but this time you have to work out the relationship between the 3 circles.

Aaaaand, go!

Jigsaw time

Take a look at these beauties below.

The idea is pretty simple, but don't be fooled — they're as tricky as they are fun.

Your task is to try to fit the yellow jigsaw pieces into each of the 3 puzzles. The rules are simple: match each number to its twin. You can also rotate the pieces to create a match. Note that there are no reflections.

I've started you off with the one below to get you into the groove. In this one, there aren't even any rotations. But don't get too relaxed: they get harder, fast.

Oh, one more thing. Can you figure out why I've placed this puzzle under number '29'?

Hop to it!

1	0				
0	1				
		1	1		
		0	1		
				1	2
				1	1

029

A cubic kilometre of seawater contains roughly 30 *million* tonnes of salt

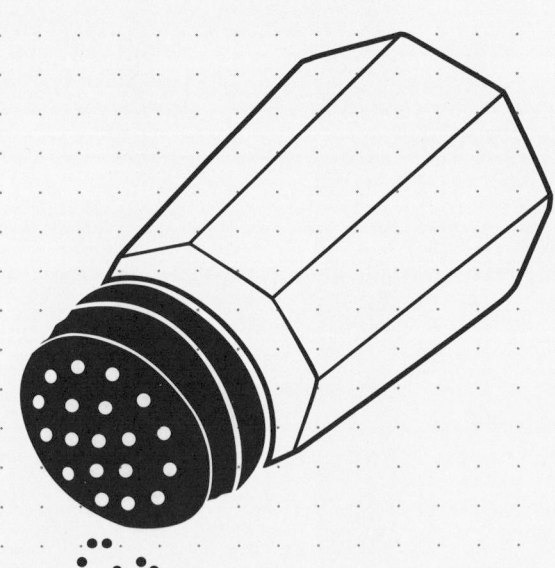

Give or take, and depending on exactly where in the ocean you are.

Dollar Detective

Sergio sells his apples at 3 for $1.

Maria rates her apples a bit more, so she sells them at 2 for $1.

One day it's getting late and they've been going since dawn. They each have 30 apples left, but really want to go home. So, they give all 60 apples to Angie and say 'just combine our prices and sell them at 5 for $2'.

Angie sells all 60 apples and earns $24 in total.

But if they'd sold them separately, Sergio would have earned $10 and Maria $15 — a grand total of $25.

Where has the missing dollar gone? And in what circumstances would there be no missing dollar?

You should expect a miracle to happen to you every 30 days

'Littlewood's Law of Miracles' says that if you hear, see or think one thing per second, then in an 8-hour day you will experience 60 × 60 × 8 = 28,800 — let's call it 30,000 — events per day. So every 30-day month you will experience about 1 million events (strictly it's 864,000). So a 'one in a million' miracle should happen to you once. A month.

It's not a hard and fast calculation, but it is a way to help convince someone who thinks they've seen a sign or are horribly cursed just to calm down a bit and roll with things.

Fill in the blanks

In each of the 30-square grids below, a list of numbers has been written down along a certain path through the grid.

The list of numbers is different each time, as is the path the numbers take as they move through the grid.

Your task? Determine the list of numbers, the path they follow through the grid and hence the number that should replace the question mark in each example.

1			7		11
			17		13
			31		35
	45			39	37
					?

1	3	4	10	11	20
2	5				21
6					27
7			23		
15	16			29	?

?	41				
		5	7	11	61
	31	3	2	13	
		23		17	71
101				79	

A billionaire at 31?

Well, there are 60 seconds in a minute, 60 × 60 = 3600 seconds in an hour and 24 × 60 × 60 = 86,400 seconds in a day. So 11 days is 950,400 seconds, leaving us 49,600 seconds to hit that magic million. Thirteen hours is 13 × 3600 = 46,800 seconds, leaving us 2800 seconds short. And 46 minutes is 46 × 60 = 2760 seconds, leaving 40 seconds to get us to 1 million.

So from the moment you were born, you turned 1 million seconds old 11 days, 13 hours 46 minutes and 40 seconds later.

I know, I know, the next question is 'when do I turn 1 BILLION (seconds old)?'. Doing the same type of calculations tells us that 1 billion seconds is 11,574 days, 1 hour, 46 minutes and 40 seconds.

But from here it gets a little bit tricky. Because 11,574 days is 31 lots of 365 days (a year) with 259 days left over. But depending on what year you were born, some of the years you've been alive will be leap years. This will change from person to person but some of you will have picked up one extra leap year over others. Now, my fellow geeks, how technical do you want to get?

We've even complicated things more by adding the occasional leap SECOND to the international calendar.

Leap seconds occur because the Earth doesn't rotate at exactly the same rate all the time. Small changes in the combined gravitational pull of the Sun, Moon and other planets, or events on the Earth itself, for example, ocean flow, variations in the molten core or even subtle changes in the shape of Earth itself(!) mean the time for the Earth to do one rotation on its axis isn't always exactly 24 hours zero seconds. Over time as satellites tell us that the Earth has fallen behind on its rotations an extra second has been added to our international timing instruments.

For example, on 31 December 2016 as we ticked over into 2017, international clocks counted 23:59:58, 23:59:59, 23:59:60, 00:00:00, 00:00:01 and off we went. Happy New Year!

A coin toss a day works your logic okay

Every day in the month of May, I've chosen to wake up this way
I'll toss a coin and if it's a head, I will get straight out of bed
But if that toss comes up a tail, I will feel that that's a fail
I'll toss again and if that's a head, then I will get straight out of bed
But if I toss a second tail, I will not gnash, I will not wail
I'll toss again and if that's a head, then I will get straight out of bed
But I will not get out of bed, until I've tossed that day's first head
So if I toss tail tail tail tail, tail tail tail tail, tail tail tail tail
Tail tail tail tail, then tail tail head, only then will I rise from bed
When I reach the end of May and I rise from bed on that final day
How many times, can you work out, should I have tossed this coin about?
(Not just on that final day, but across the entire month of May)

Here's a hint.

The mathematics required to show the answer here might be a bit hard for some of you. If that's the case, why not grab a coin and simulate a 10-day month and count how many tosses in total you needed.

Try this for 10 or so 10-day months and what looks to be the answer? So what answer does this suggest for the 31 day month of May? Convince yourself why that answer makes sense without trying any calculations.

031

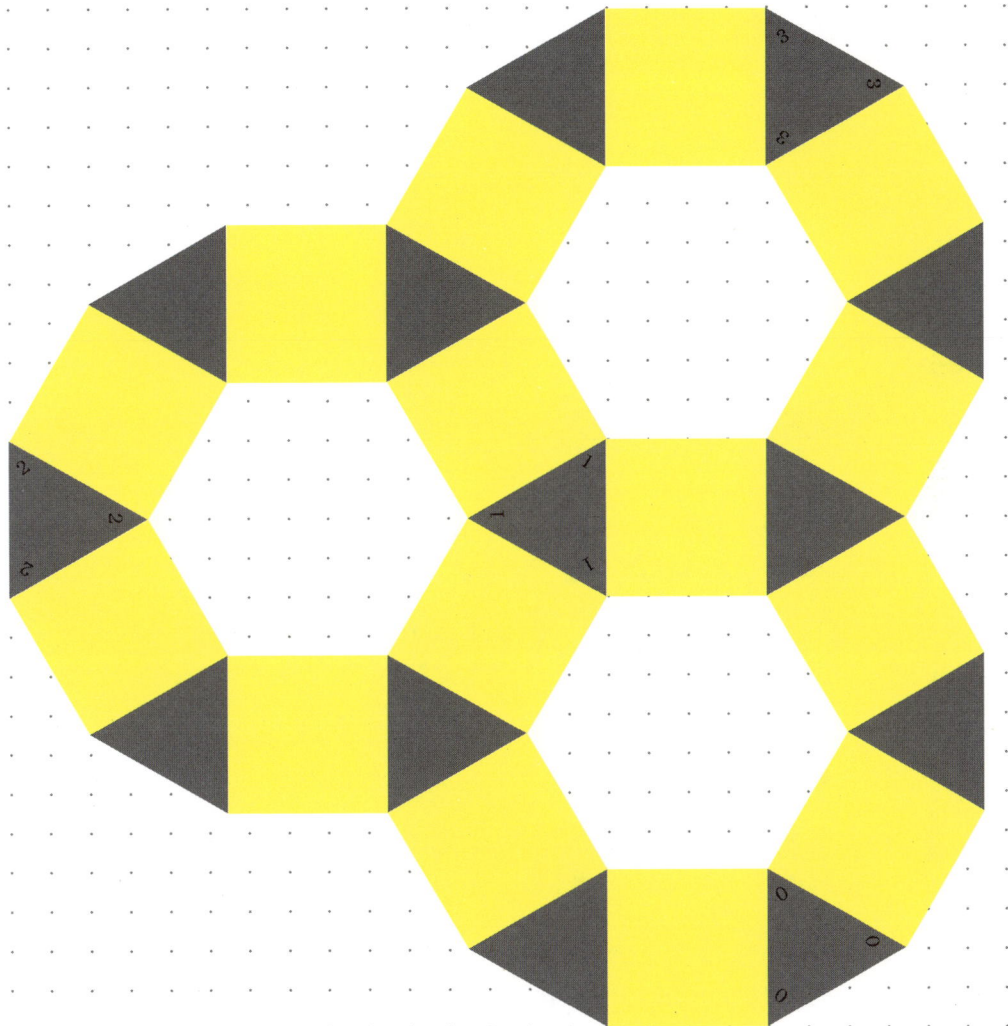

Triangle and square (like you just don't care)

You know the drill by now.

Rotate (where necessary!) and place the pieces on the opposite page on this beauty above. Remember, the numbers must match (0s with 0s, 2s with 2, and so on).

For bonus points, tell me – why is this puzzle in '31'?

Adults typically have 32 teeth

Unless they've had their wisdom teeth removed or list 'UFC fighter' as their occupation on their passport.

Looking after our teeth is a lot easier these days if you have access to fluoridated water and brush and floss regularly. No matter how much of a chore caring for your teeth may seem at times, be thankful you're living today and not in the Middle Ages when it was thought toothache may be caused by tiny worms in your teeth. The Middle Ages gave us this often quoted remedy:

'Take a candle of sheep suet, some eringo seed being mixed therewith, and burn it as near the tooth as possible, some cold water being held under the candle. The worms (destroying the tooth) will drop into the water, in order to escape from the heat of the candle.'

I presume I don't even need to explain that 'sheep suet' refers to the layer of fat found around a sheep's kidneys for you to agree, yeah modern dentistry is a very good thing.

___ × ___ × ___ ÷ ___ + ___ + ___ + ___ = 32

___ × ___ − ___ × ___ − ___ ÷ ___ + ___ = 32

___ × ___ − ___ × ___ × (___ − ___ + ___) = 32

((___ × ___ + ___) ÷ ___ − ___) × ___ − ___ = 32

((___ + ___) × (___ + ___) + ___) ÷ (___ − ___) = 32

Your time starts ...

Now! The answer 32 uses the digits 2 and 3. Complete the equations using the other digits 1, 4, 5, 6, 7, 8 and 9 exactly once each.

Go for it!

Big raindrops hit the ground (or your head) at about 32 kilometres per hour

Smaller drops reach the Earth at a far lesser terminal velocity of about 7 kilometres per second.

$$1 = \frac{4 \times 4}{4 \times 4}$$

$$2 = \frac{4}{4} + \frac{4}{4}$$

$$3 = \sqrt{4} + \frac{4}{4}$$

$$4 = (4 - 4) \times 4 + 4$$

The four 4s

Have a look at the 4 equations above.

One famous, fun and very frustrating maths problems is this: with only basic mathematics, write all of the numbers from 1 to 100 using exactly four 4s.

When we say 'basic mathematics', you only need +, −, ×, ÷ along with ! (see chapter 008), $\sqrt{}$ ($\sqrt{4}$ = 2) to get most of the numbers — and a couple more obscure signs to work out the toughest ones.

Using the basics, including 4! = 24, $\sqrt{4}$ = 2 and using two 4s to get 44, you can make all the numbers up to 32 with up to four 4s, without too much trouble. Well, I tell a bit of a fib —the answer for 31 is a beast. Anyway, give it a go!

A solar mass is a unit of mass used to measure stars

It's based on the mass of our Sun. To work out a solar mass we need to know a few things. Firstly, 1 AU is 1 'astronomical unit' or the average distance of the Earth from the Sun, about 149,597,870.7 kilometres. Secondly, G is the gravitational constant, an amount we use when calculating things to do with gravity. It's approximately equal to:

$$G = \frac{6.67408 \text{ m}^3}{10^{11}\,\text{kg}^2\,\text{s}^2}$$

We just whack these amounts into the handy formula:

$$M_{\odot} = \frac{4\pi^2 \times (1\,\text{AU})^3}{G \times (1\,\text{yr})^2}$$

and we get $M_{\odot}$ = (1.98855±0.00025) × 10^{33} grams = 1,990,000,000,000,000,000,000 ,000,000,000,000 grams, approximately.

It is conventionally written in kilograms, not grams, as $M_{\odot}$ = (1.99 ×10^{30} kg) which makes the Sun sound just a little bit slimmer. Regardless of how you write it, the Sun is 332,946 times the mass of Earth, 27,068,510 times the mass of the Moon or 1048 times the mass of Jupiter.

You can just take my word for this, or if you've got a scientific calculator and a handy table of planet masses you can actually work it out. Not for everyone, but true maths nerds, have a crack.

Now 1000 grams is a kilogram, and 1000 kilograms is a tonne. So the fancy maths talk for a solar mass is give or take, 2000 yottatonnes!

Take a hike!

You're standing on the surface of the Earth. You walk 33 kilometres south, 33 kilometres west, and 33 kilometres north.

You end up exactly where you started.

Where are you?

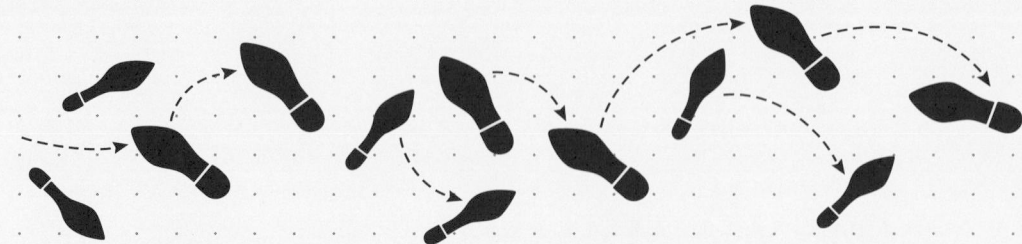

500 years ago, over a week in July ...

34 people joined Mrs Troffea who was dancing wildly in Strasbourg, France and the 'dancing plague' of 1518 was born.

Within a month, over 400 people, mostly women, had joined the mad gyrations and some had danced until they died of heart attack, stroke or sheer exhaustion. Rather than stop them, the local authorities seemed to think that letting them dance it out was the best solution. They even constructed stages and hired musicians to keep the dancers going.

Here are two theories as to why this epidemic of dancing took place. Historian John Waller, in his book A Time to Dance a Time to Die *suggests that the hardness of the times, repeated famines, disease, starvation and war led to a mass stress-induced psychosis — that people were literally dancing insane. Another theory is that a fungus growing on wheat crops caused LSD-like hallucinations and that frantic dancing ensued.*

Whatever the cause, the dancing plague lasted until September when the remaining dancers were transported to a local mountain where they were made to pray for forgiveness.

Concatenation

Here's one fun little mathematical recreation that we will visit a few times in this book.

Take the numbers 1 to 9 in numerical order and, by using just three tools — addition, subtraction and concatenation — create an equation that gives an answer of 100. Huh? What the deuce, Adam? What on Earth is concatenation? Don't panic, it's just a fancy way of saying 'joining numbers' — so that 3 and 4, for example, become 34.

We could concatenate 2 with 3 to get 23 and 7 with 8 to get 78 and the equation:

$$1 + 23 - 4 + 5 + 6 + 78 - 9 = 100$$

In fact, 23 pops up again in:

$$1 + 23 - 4 + 56 + 7 + 8 + 9 = 100.$$

All told, there are 12 equations like this. They all begin with positive 1, except for one example.

Only one of them involves 34. I'd like you to try to find it. If we just add the remaining numbers 1, 2, 5, 6, 7, 8 and 9, we only get 38 so, obviously, you need at least one more concatenation.

In fact, you only need one. A good strategy here is to start by considering the various concatenations possible and trying to get the remaining amount you need out of the other digits. There is also a very nice observation you can make about whether the other concatenation is odd or even and the effect it has on the rest of the sum that you can obtain from the other digits! Good luck.

035

Typical seawater is around 35 parts of salt per 1000 parts of water

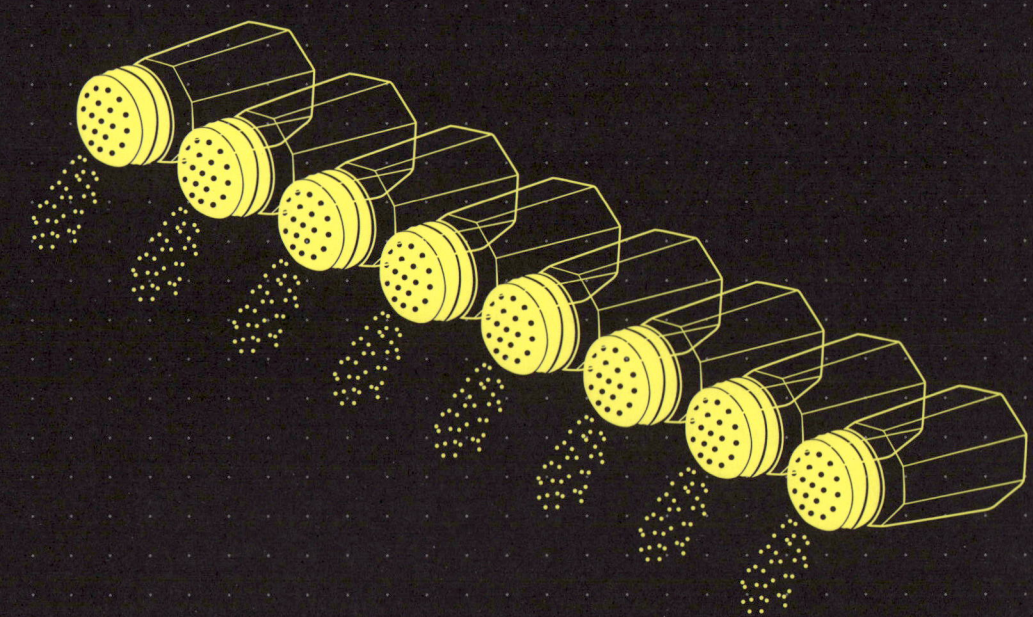

Or to put it another way, the salt comprises about 3.5% of the seawater.

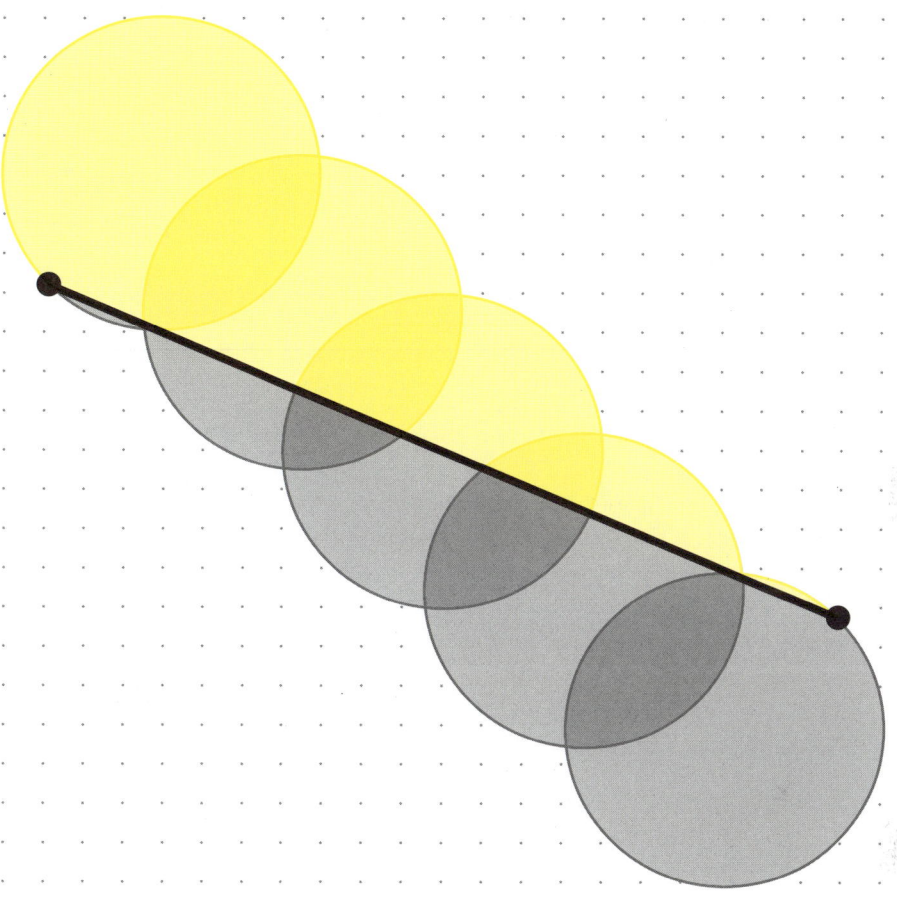

Going around in circles

Check out the 5 circles above.

Notice that where they overlap, they create 4 regions. Each of these overlapping regions of two circles has an area of 5, and the entire yellow section has an area of 35.

Your task, should you choose to accept it (and, hey, why wouldn't you?) is to figure out the area of one of the circles.

Roulette wheels have 36 red and black numbered slots...

Along with a 0. There is a 00 slot also in use in casinos in the United States, Canada, South America, and the Caribbean.

So for every spin of the wheel in a 0-only casino, the 0 should come up 1 in every 37 spins. Another way of thinking of this is that if you take a $1 coin and place it onto one of the numbers 1 to 36 you will win 1 out of every 37 spins. You get paid $35 for each successful spin. You will lose the other 36 out of 37 times when the 0 or any number other than yours comes up.

So the 'expected' result for a given spin with a $1 coin on a single number is:

$(-\$1) \times 36/37 + (\$35) \times 1/37 = -\$1/37 = -\$0.0270$ ... giving the house an 'edge' of 2.7 cents in every dollar laid down.

Now this is very different to saying that the house only collects 2.7% of the money bet at a casino. This is because if you start with $100 and lose 2.7% of your money, you're down to $97.30. If you gamble that $97.30 and lose 2.7% of it, you'll have $94.67, and if you keep gambling that you'll lose more and more.

You can easily lose all of your money playing a game that has a house edge of just 2.7%, though you could win and leave or stop playing when you've only lost some of your money. Reporting by the Casino Control Commission in Atlantic City in the US suggests casinos running double zero roulette wheels average a hold of 21–30%, which is the amount of money a player typically loses to the house per session.

The lesson here is this: if you choose to gamble, assume you will lose and please only gamble what you're willing to say goodbye to.

Three 36 thinkers to think through thoroughly

Here are three short ones for you.

Thirty-six people go on a camping expedition. They have enough food for 4 days. Due to a mix-up, an extra 12 people turn up at the campsite. How long can they now stay before the food runs out?

Once you're back from your camping trip, have a crack at this one.

It takes 9 women 36 days to build a house. How long would it take 12 women?

If you only needed the house built in 81 days, how many women would you need?

I'll give you bonus points if you can say the title of this puzzler 10 times fast while standing on your head!

Dinner at 37

Puzzle Place isn't just a great place to take your brain for a wander. It's also got some of the best restaurants in town. So good are these nosheries and so broad is the range of cuisine that every 3 months, 4 couples celebrate the new season with a dinner.

In summer they eat a seafood barbecue, in autumn bring on the pizzas, in winter it's piping hot Indian curry and in spring they prefer sushi.

As you pass through this book you'll come to love these 8 crazy cuisine consumers as you wrestle with the common problem of working out the way they sit around the table for that particular meal. The restaurants all have circular tables and the seats are numbered clockwise from 1 to 8. Aaron organises the lunch and always sits in seat number 1 before then telling everyone else exactly where to sit (control freak).

Let's join them now for their summer meal at Simone's Sizzling Seafood, situated at 37 Puzzle Place. The 8 people sitting at the table, in no particular order, are Aaron, Barry, Caroline, Danielle, Eamon, Fantine, Gary and Hayley.

For this meal we know that:

- Aaron sat in seat number 1 opposite Eamon.
- No two neighbours had names that shared a last letter.
- No two neighbours both had names with an odd number of letters.
- Caroline was the only person whose name had the same third letter as the person to their immediate left.

Can you write out the seating arrangement for this sumptuous summer snack?

Chromosomes are amazing collections of genetic material

Humans have 46 chromosomes which occur in 23 pairs and every cell in our body has a copy of our chromosomes floating around in its nucleus. Strictly this isn't true because male sperm have only 23 chromosomes as do female eggs and part of the magic of a man and a woman liking each other very much is when they unite, their 'sex cells' to give the baby a different version of 46 chromosomes to either Mum or Dad.

Interestingly, from around 1923 until 22 December 1955 it was thought we had 48 chromosomes. This was based on a faulty reading through a microscope and remained the case until late '55 when the great Joe Hin Tjio set us straight. Nice one, Joe.

Having 38 chromosomes in 19 pairs is very trendy in the animal and plant kingdom. Just ask any pig, cat, tiger, lion or grape. They may struggle to answer, especially the grape, but they all have 38 chromosomes.

By the way, check out @biolojical on Twitter, who teaches biology via his feed ... using emojis.

Boy germs!

Ms Johnson is taking 38 students out to lunch.

They are sitting in the café area at tables that can take from 4 to 6 people. Now, Ms Johnson has been around the block a few times and knows that, as much as possible, she should keep the boys and girls separate when they're eating.

If she sits the students 4 to a table, 9 tables will be full and one will have 2 boys left over.

If she sits them 6 to a table there will be 6 full tables and one with 2 girls left over.

If the class contains slightly more girls than boys, how many girls are in Ms Johnson's group?

Mathematicians have all sorts of whacky names for different sorts of numbers

39

Even numbers (whole numbers that are divisible by 2), prime numbers (whole numbers not divisible by anything other than 1 and themselves), transcendental numbers (trust me, really cool) are just a few examples.

Well, there's even a class of numbers called 'evil' numbers! A number is 'evil' if when you write it out in binary, or base 2, you need an even number of 1s.

The binary expansion of 39 is 100111 which requires four 1s, proving the 'evilness' of 39.

And numbers that are not evil? They're called 'odious', of course.

One the topic of evil, read on ...

Evil non-genius

We've all been there.

You're captured by your evil archnemesis and taken into a room.

There is a massive table in the far corner and you can see it is covered in gold coins. You're too far away to see how many coins, but it must be hundreds, maybe even thousands. Like I said, it's a really big table.

You know how this works. If you've ever seen any action adventure film, from James Bond to Austin Powers, you know your evil nemesis isn't just going to shoot you dead on the spot. No, no, no. They will have some complicated evil plan to enjoy watching you suffer. But *this* complicated evil plan is also your ticket out of here.

'Ha, ha,' she laughs, blindfolding you, walking you across the room and seating you in a chair at the table. 'I won't tell you how many coins are on the table, but I will tell you this. They are all solid gold and worth in total over a million dollars. And each coin has "live" engraved on one side and "die" on the other. I know that 39 is your lucky number, so I've arranged them so that at the moment exactly 39 of them show "live" face-up and the rest all say "die".'

She pauses for dramatic effect.

'Your challenge, whether you choose it or not, is this. You can move the coins and flip as many over as you like. But you can't look at them. And you'll realise very soon that no amount of feeling the coins will help you realise which say "live" and which say "die".

'If you can split the coins into two groups, of any size, with exactly the same number of coins showing "live", I will let you go. But if your two groups do not have exactly the same number of coins showing "live" ... you guessed it ... you will die! Mwaha mwahaha mwahahahahahahahahaha!

'Okay, you've got one minute ... best of luck.'

The clock is ticking! How do you produce two piles with exactly the same number of coins that say 'live'?

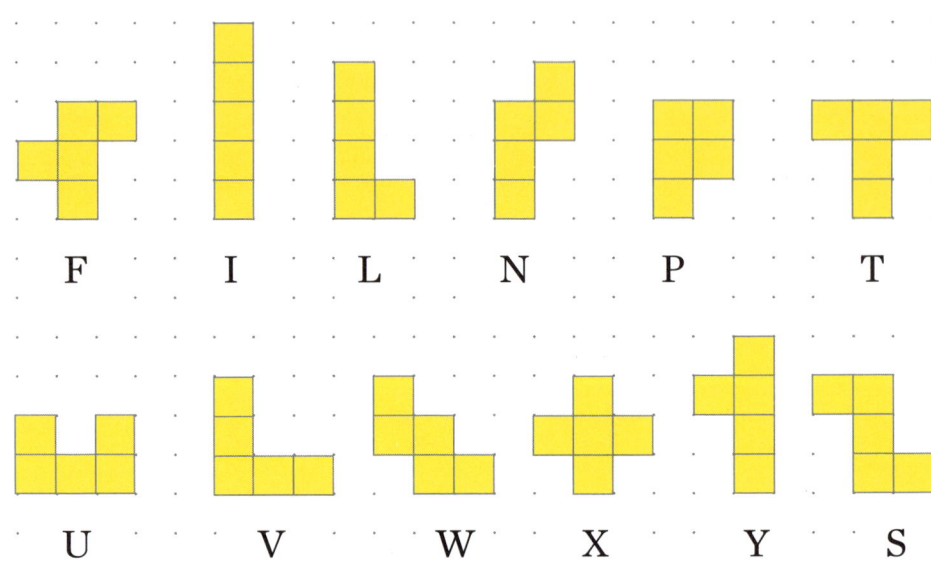

F	I	L	N	P	T

U	V	W	X	Y	S

STIX it to me!

You'll be quite familiar with the free pentominoes by now. But here's a fun twist.

Once again we have the 12 pentominoes — joining 5 squares along their edges. Remember that any others you make are just reflections or rotations of these.

One way to describe the pentominoes is by the letters they (okay, sort of — use your imagination) look like, though I've used S when most mathematicians flip it around and call it Z.

In this puzzle, which I've called 'STIX it to me', I want you to place the remaining 8 pentominoes to fill the grid. This also explains why I went with S, not Z. 'ZTIX it to me' is even lamer than 'STIX it to me'.

For bonus points, one of the diagrams actually has 2 possible solutions. Can you find both of them?

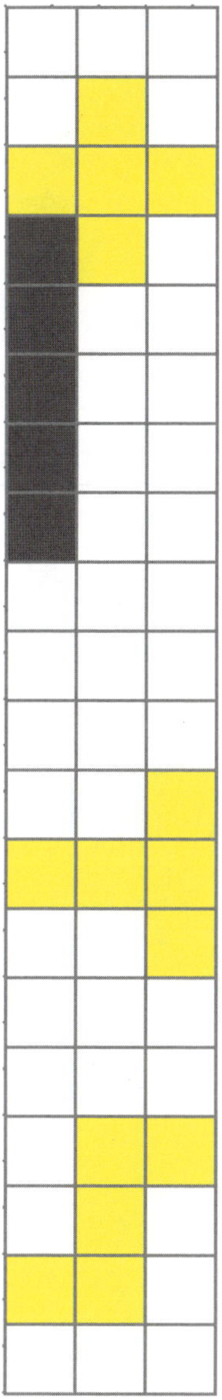

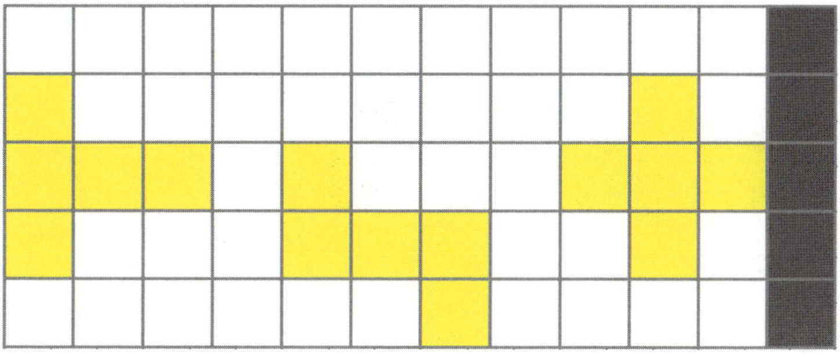

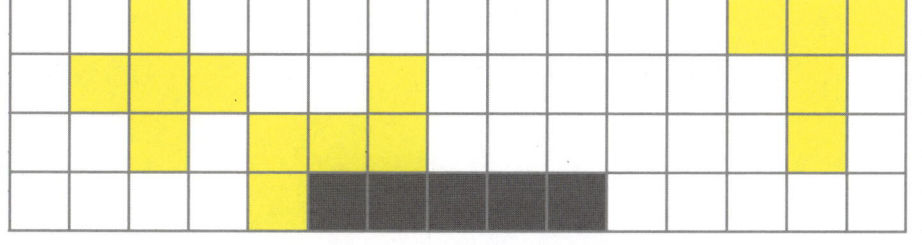

It's estimated that up to 40% of the Vietnamese population have the surname Nguyen

In total, the 14 most popular surnames, Nguyen, Tran, Le, Pham, Huynh, Phan, Vu, Dang, Bui, Do, Ho, Ngo, Duong and Ly cover perhaps 90% of all Vietnamese! In the US the 14 most popular surnames cover more like 7%.

The top 3 names also happen to be names of Emperors and dynasties which tends to make you just a little bit popular with the common folk!

The longest earthworm ever dug up in the United Kingdom was 40 centimetres long

It was discovered by Paul Rees. Paul's stepson George named the worm Dave, as opposed to its scientific name Lumbricus terrestris.

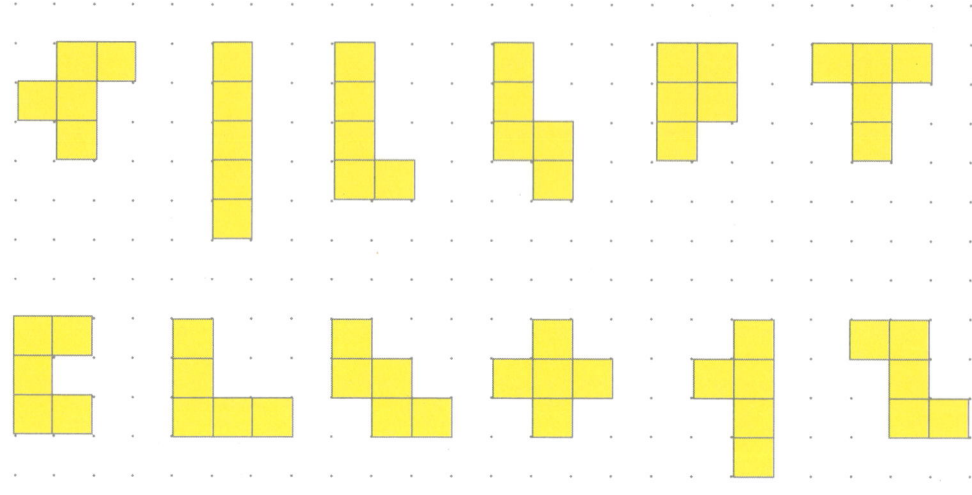

Pesky pentominoes

Yes those pretty but pesky pentominoes are back. One of the questions for the number 5 introduced us to the 12 free pentominoes.

Well, now I've removed the centre square from the 41-square grid on the opposite page. The remaining 40 squares contain 8 sets of the letters ABCDE each filling up a pentomino shape. For example, you might find this ...

	B	D
E	A	
	C	

... somewhere in the grid.

The challenge is to find 8 interlocking pentominoes so that every one contains all the letters ABCDE. You can use a pentomino twice, or rotate and reflect the free pentominoes as given.

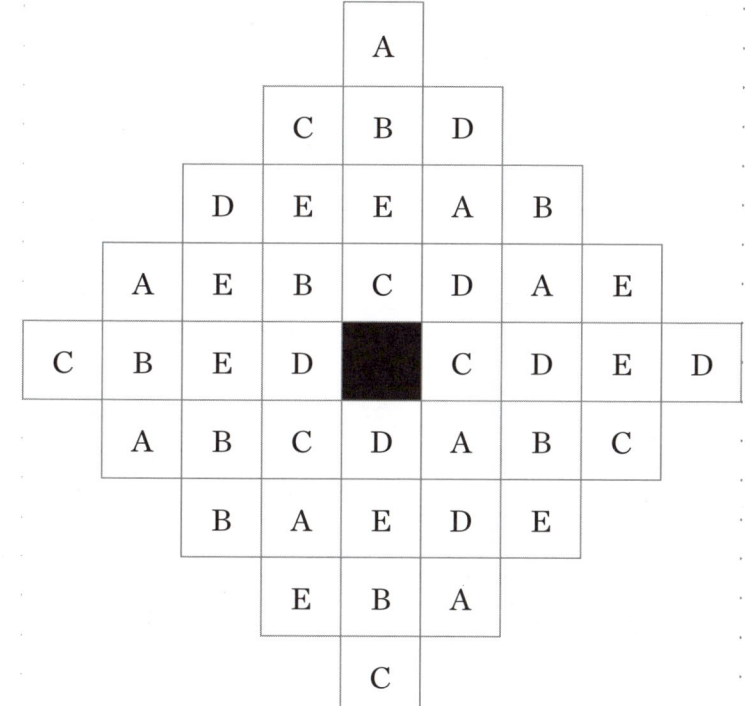

Here's a hint: start at the corners and see if any pieces are 'forced' — that is, there is only one possible pentomino that can contain that corner square.

Get to it!

Mozart's 41st Symphony ...

(sometimes known as the Jupiter Symphony) was his longest ... and his last.

If you'd like to play it with a few mates, you'll need access to a full strings section, a friend who can play the flute, another who can play timpani, and two people each who can play oboe, bassoon, horns, trumpets.

You'll also need roughly 33 minutes.

6	2	8	7	9	7	0
7	4	3	5	5	7	7
5	3	3	5	10	10	10
4	9	12	8	12	10	11
1	6	9	12	9	9	11
9	6	5	11	10	11	9
9	6	6	11	9	10	11

15	15	5	6	5	5
5	6	6	4	6	4
4	5	5	5	6	6
6	6	5	5	5	5
5	6	6	4	4	4
5	4	6	5	6	5

Forty-one fun

Draw 4 straight horizontal or vertical lines through the top grid above so that the numbers in each section add up to exactly ... you guessed it, 41.

No surprise that the second one of these is much tougher. Draw 3 lines through this grid, creating 5 sections that contain numbers that add up to 41.

How do we only get 5 sections? Well, two of the lines run the length of 6 squares from side to side or top to bottom, but the third line is only 4 squares long so starts at one edge of the grid but doesn't make it all the way to the other edge — it stops when it hits one of the other lines you've already drawn! Ouch!

Water freezes at 0° Celsius, yeah? Well, sort of

For water to freeze, the molecules of H_2O need to be able to 'freeze onto' something — impurities like small particles of dust or 'nucleation centres' caused by the water molecules vibrating. For normal tap water, there is enough roughage and vibration in it for freezing to take place at 0 degrees.

But if water is absolutely pure and kept very still, its temperature can be pushed down to –42 degrees Celsius before it will freeze.

If you had a container of this 'supercooled' water and you hit the side, the vibration introduced to the molecules would see it instantly freeze. We sometimes see this with rain drops that freeze as they hit the earth.

To further complicate things, salt performs the opposite role to dust. Salt actually slows down the freezing process. So when British man Lewis Pugh breaks world records by swimming in water that is 1.7 degrees BELOW zero (in only SPEEDOS!) he is doing so in ocean water.

In fact, if water is incredibly salty, say 23% or more, you can get it down to –21 degrees Celsius before it will freeze.

Brrrrr!

1	1	3	4	2	2	4	4
2	1	4	1	1	3	4	3
1	4	4	2	4	2	4	2
4	1	3	3	1	1	3	3
4	3	1	3	1	3	2	3
4	3	4	2	2	2	4	1
3	1	3	1	4	4	3	3
4	2	3	3	1	3	2	4

1	2	4	4	1	2	1	2
4	4	1	4	3	3	2	2
2	2	1	1	4	4	2	1
2	4	3	2	4	4	3	2
4	4	2	3	1	3	2	4
1	4	3	3	2	3	2	4
3	3	1	4	1	4	1	2
2	3	1	4	3	4	2	4

Life, the universe and these brainbusters

You might have come across those 'divvy up the plot of land' puzzles in the past.

Well, me being me, I've added a numeric element here for your delight.

Your goal is to divide each grid into 4 identical shapes of 16 squares each. Not so fast! The sum of their interior numbers must equal our target number — 42. The identical shapes themselves are unusual, so hard to find. Good luck!

Female duck-billed platypuses grow to an average of 43 centimetres ... and *rock*!

The glorious duck-billed platypus is Australia's own and one of the most fascinating animals in the world. So beautiful but bizarre looking that when it was first presented some scientists thought it was a 'hoax' created by gluing bits of various animals together. This sounds weird but it was a popular trick by naturalists at the time to create 'hybrids' and see who they could fool.

Platypuses (along with a few species of echidnas) are monotremes, or mammals which lay eggs. Females lay typically 2 eggs after a 21-day gestation period and the young feed off mummy (who, lacking teats, sweats milk) for around 4 months before emerging from the burrow. They can then live for up to 17 years.

The males have a long spur on each ankle which produces a venom that while not lethal, really hurts according to all accounts. If you ever have to pick up a platypus, it's advised to do so holding the end half of the tail and keeping a close eye on their feet.

Measuring its REM sleep, the platypus seems to be the animal that dreams the most, maybe 4 times as much as you or me.

One final freaky food thing, a platypus has no stomach and can store up to 600 worms in the pouches of its cheeks.

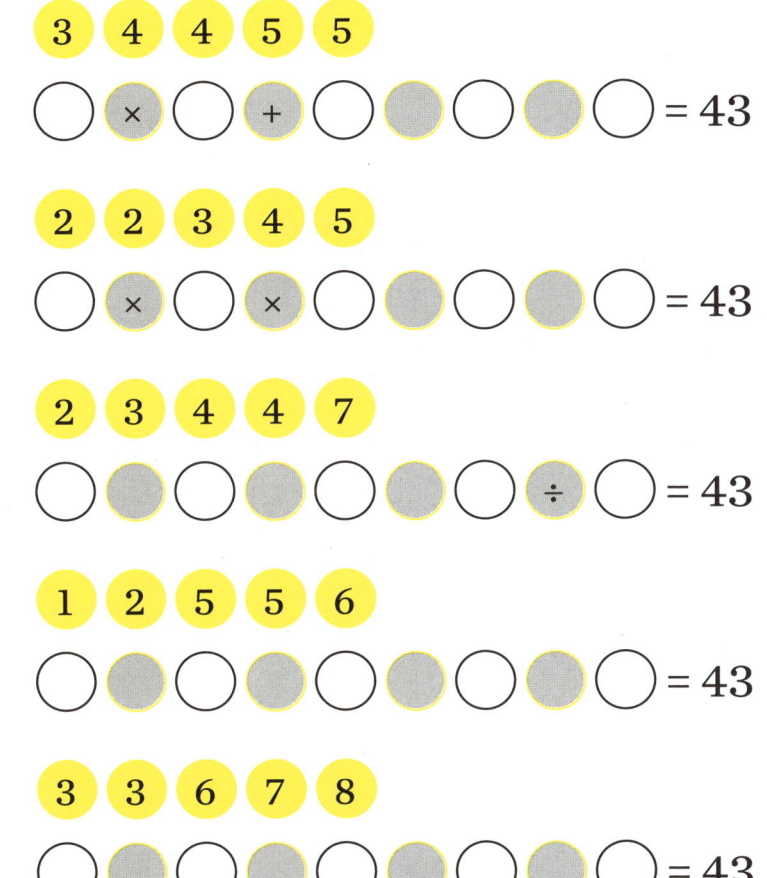

3 4 4 5 5

◯ × ◯ + ◯ ◯ ◯ = 43

2 2 3 4 5

◯ × ◯ × ◯ ◯ ◯ = 43

2 3 4 4 7

◯ ◯ ◯ ◯ ÷ ◯ = 43

1 2 5 5 6

◯ ◯ ◯ ◯ ◯ = 43

3 3 6 7 8

◯ ◯ ◯ ◯ ◯ = 43

Forty-three fums

So you've slayed the sums at 17?
Well, why not make a fist of figuring these forty-three, er, fums?

Same deal here: find where the number and the operators fit to hit our target number, 43. Don't forget that you're trying to find the arrangement that gives you the highest 'score', found by reading off the yellow numbers in order.

You've likely heard of the Dead Sea which, at 33.7% salt, is 10 times as salty as typical seawater

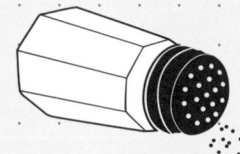

But there is a saltier place ... much saltier.

Come on down the Don Juan Pond, a body of water barely 10 centimetres deep in the McMurdo Dry Valleys of Antarctica which is a whopping 44% salt! There is so much salt in DJP that the temperature would have to get to –53° Celsius for it to freeze. Well, it's a balmy –25° or so most days in the valley so the pond stays liquid ... and I'd imagine, very 'refreshing'.

Thick as a brick

Georgio can lay 44 bricks in an hour.

His sister Esmerelda, who is new to the trade, can lay 22 bricks in an hour.

How long would it take them to lay an 88-brick wall?

Too easy? Well, their pesky little brother Luigi walks along behind them taking down 11 bricks an hour.

How long does the 88-brick wall now take?

In 1984 16-year-old video game whiz Tim McVey became the first to score over 1,000,000,000 points on a video game

He did it on an otherwise unspectacular game Nibbler which was one of the few games programmed to allow scores of this size. The epic feat took 44 hours and occurred at the legendary Twin Galaxies video arcade in Iowa.

Twenty-five years later, to end decades of controversy surrounding other players who claimed to have broken the record, McVey went back again and crushed one billion on Nibbler. This is all recorded in the excellent documentary Man vs Snake.

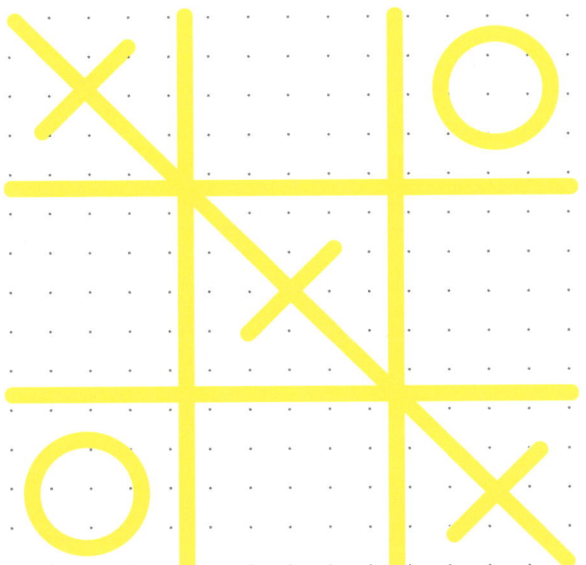

O's and X's ii

Public service announcement!
You might be best to try the noughts and crosses question at 091 before doubling back to here — this one is actually a fair bit tougher.

From 091 we've seen that there are 1440 possible games that end with X winning on the 5th move (spoiler alert, if you didn't trundle over to 91 as a result of my PSA).

There are only 44 configurations in which O can win in noughts and crosses, but the quickest way it can win is on the 6th move.

There are 5760 – 432 = 5328 possible 6-move games. Prove it!

Hint: the method to use here is a common one for more difficult counting problems which we touched on back at the flag problem in 004. First of all, work out all the possible grids three Xs and three Os where the three Os are in a winning line. Then, subtract any grids in which the Xs are also in a winning line, because in this case X would have already won. This is where we get 5760 – 432 = 5328 from. Okay? Get cracking!

Pentominomenal!

Way back at number 5 we met the 12 free pentominoes. Gorgeous shapes constructed by joining 5 squares along their edges. Hopefully you managed to discover the 12 free pentominoes (if not, spoiler alert):

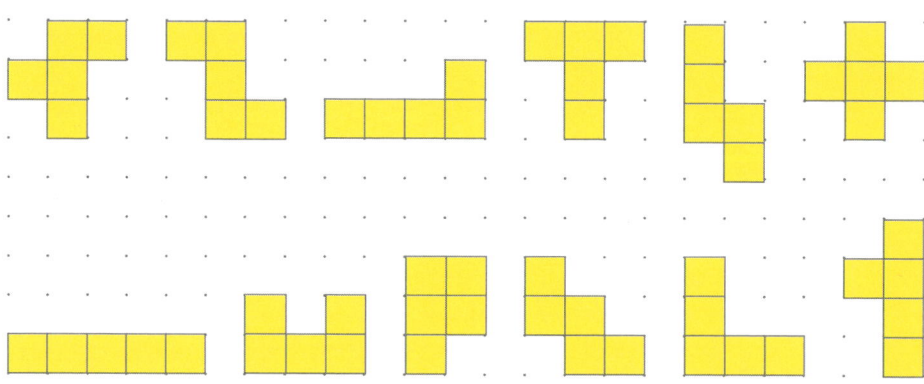

As we saw back at 040, one of the many awesome things about pentominoes is that they can fit together into beautiful shapes. Let's explore this further.

Twelve pentominoes made up of 5 squares each makes 60 squares and it turns out that the 12 pentominoes can be fitted together to cover the 60 squares of a 6 × 10 rectangle. In fact, they can do this 2339 different ways.

They can also be arranged to fill a rectangle of dimensions 5 × 12 (1010 ways) or 4 × 15 (368 ways) or 3 × 20 (for which there are only 2 possible solutions).

You might have noticed that in each of the grids on the opposite page, I've already placed 3 pentominoes.

Can you fill the remaining 45 squares with the 9 free pentominoes I haven't yet used? It will be interesting to see if you find this much harder than the 040 puzzles.

I've started with the 3 × 20 because it's the easiest. Don't forget you can go to adamspencer.com.au to download a pdf of this puzzle if you'd like to get messy before getting your final answer.

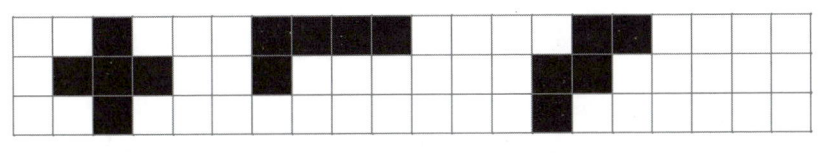

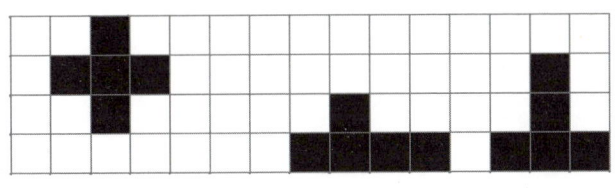

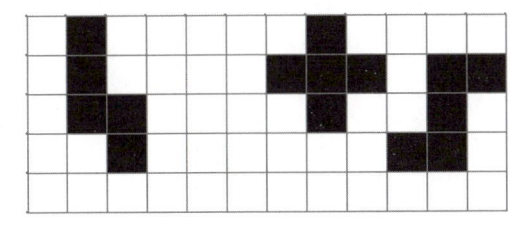

There are up to 45,000 different English surnames

So says the family tree website Ancestry.com, and the vast bulk of them belong to one of 7 categories. The name tells you something about your ancestors, for example what they did for a living or where they grew up.

The 'families' of names are:

1. *Personal characteristics (names like Short, Long, Little ... even Peacock in the case of someone who was a little bit up themselves!).*
2. *English place names (one of the most famous English surnames of today, the 'Beckhams', may well have come from Beckham, in Norfolk).*
3. *Occupations (when you think about it, a lot of people you know have surnames that are occupations. These might be obvious in the case of Archer, Baker, Cook or Gardener, but many occupational surnames go by unnoticed these days because people don't do that job any more. Think of your friends who are named Cooper [a barrel maker], Faulkner [a trainer and keeper of falcons!], Smith [the village blacksmith] or Wright [a woodwright was a shaper of wood]. In fact, I never knew that the surname Spencer came from the court position of dispenser or steward, an official who was in charge of certain parts of the land!)*
4. *Names of a castle or estate (like the British Royal Family Windsor).*
5. *Geographical location names (think Forest, Hill or Wood).*
6. *Ancestral surnames like Jackson (son of Jack) or Marriott (from Mary). These sorts of surnames are common in lots of cultures. If an Icelandic man named Jón has a daughter, she carries the name Jónsdóttir.*
7. *Surnames are those that honour a patron (for example, if your surname was Kilpatrick it didn't mean you wanted to kill a guy named Patrick. Far from it. It actually meant you rated Saint Patrick so highly your family were devoted followers of him!). Who knew?*

9	6	8	3	**23**
5	7	1	2	**14**
8	3	1	3	**29**
2	4	1	5	**21**
20	**19**	**27**	**21**	

Gridlock

In the grid above, you add all the numbers in a row or column to get the answer at the end of that row or column.

One small problem, which you may very well have spotted: while the totals in yellow are correct, every single number in the 4 × 4 grid has been swapped with another. So maybe 6 should really be 3, or perhaps it represents 8? Could 7 really be a 2? The only thing you know is that none of the numbers stand for the value they represent.

Oh, you do know one other thing: $1 + 2 + 3 + 4 + 5 + 6 + 7 + 8 + 9 = 45$. Use that information to work out the correct values in the grid.

And if you're still stuck? Spoiler alert: there's an extra hint below ...

Because $1 + 2 + 3 + 4 + 5 + 6 + 7 + 8 + 9 = 45$, if you can find two rows or two columns that use 8 distinct numbers, add their totals together and the missing number is what would increase this sum to 45. Good luck.

The divisors of the number 10 are 1, 2, 5 and 10

So? Add together the divisors apart from 10 itself (the 'proper divisors' of 10) and you get 1 + 2 + 5 = 8. Now 8 is less than 10, so mathematicians call 10 a 'deficient' number. Hey 10, it's not an insult, just the way the jargon rolls.

In the case of 12, whose proper divisors are 1, 2, 3, 4 and 6, we see that 1 + 2 + 3 + 4 + 6 = 16 which is greater than 12. We call 12 'abundant' and again 10, don't take it personally.

On rare occasions like 6, the proper divisors 1 + 2 + 3 = 6 exactly. The Greeks called 6 a 'perfect' number. (Now 10 is really getting annoyed!). The next few perfects are 28, 496 and 8128.

Why am I telling you all of this? Well, 46 is the largest even integer that cannot be written as the sum of two abundant numbers.

YOU
ARE
HERE

Conundrum Crescent and Logic Lane

The 5 houses on Puzzle Place, numbered 42, 44, 46, 48 and 50, sit between cross streets Conundrum Crescent and Logic Lane.

The residents of these houses are Brenda Black, Belinda Brown, Gareth Green, Georgie Grey and Wilma White, and I can tell you the following things about the houses and the people who live in them:

1. The residents love their names so much they've actually painted their houses the matching colour.

2. Georgie Grey lives between two houses that begin with a 4.

3. No next-door neighbours have surnames that begin with the same letter.

4. Wilma White lives in a corner house, but Gareth Green does not.

5. Brown's street number is greater than Black's and White's.

So, who lives in number 46?

047

Forty-seven years worth of duty free!

When trying to plan a return flight from Christchurch to London, a British tourist was rather surprised when the Skyscanner travel planner offered him a rather long stopover in Bangkok ... of 413,786 hours ... and 25 minutes. Ah, technology.

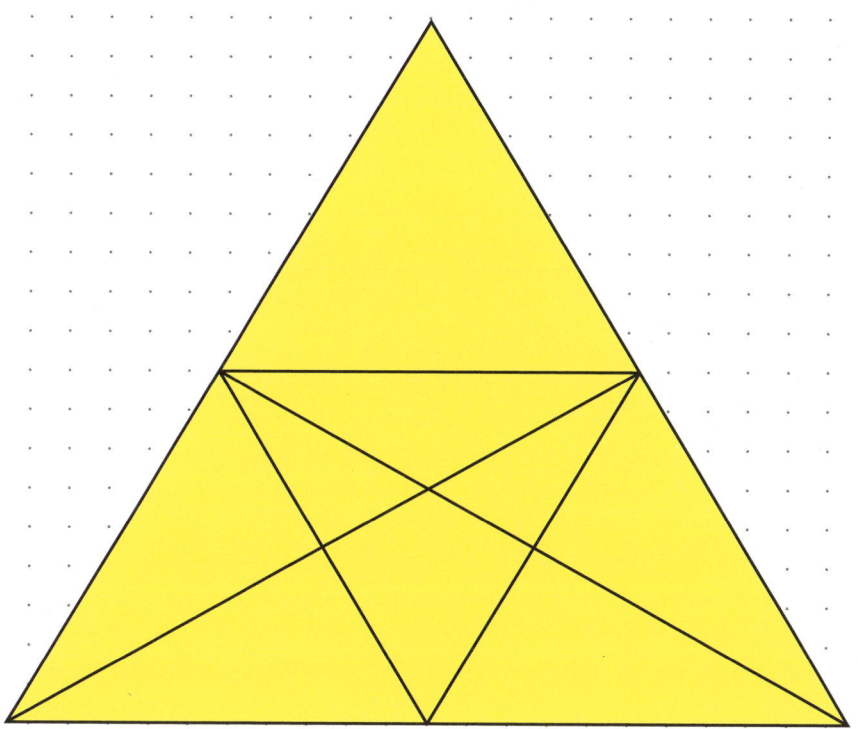

Triangle wrangle

When I was a naive first-time author I made an error or two. In particular, I included this incomplete diagram in a question in my *Big Book of Numbers*.

Of course, I put these 'errors' in just to test whether you were awake. You know, they were *completely* intentional. Well, if you were one of the keen-eyed readers who passed the test (wink!), this ought to be a cinch. You can also quietly smile knowing that the record is now clear!

First up: how many triangles can you see in the image above?

And secondly: where can you draw one single, straight line to create the magic 47 triangles I wrongly claimed could be found in this diagram?

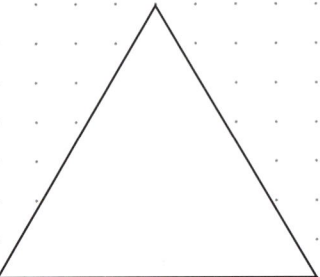

Toughen up, infinite snowflake

You may remember back at 27 how we took a simple cube, messed around with it, and came up with a gorgeous if slightly freaky object called a Menger Sponge.

Well, here's a way to mess with the humble triangle and get something similarly awesome.

Start with a triangle with all sides equal. You know, just like the one above. Bonus points if you recall that this is called an 'equilateral' triangle.

Now, break each side into thirds and in the middle third of each side build out a smaller equilateral triangle, like so:

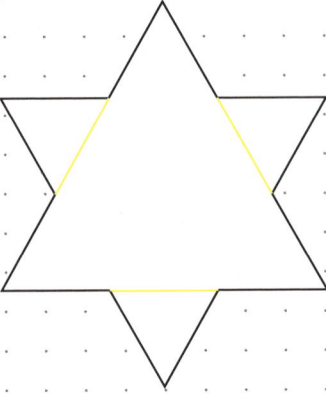

Now, do the same to each side of the new shape. Divide it into thirds and place an equilateral triangle in the middle of each side:

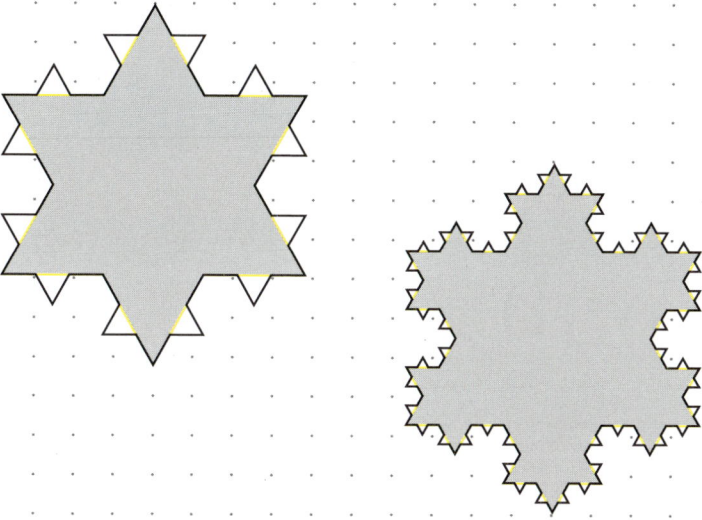

If you continue this process forever, the shape becomes what is called a 'Koch Snowflake', after Swedish mathematician Helge von Koch who first started playing around with them at the beginning of last century.

If you count the number of sides of the first few snowflakes you should see that there are 3, 12 and 48 sides. After that it is probably too hard to draw one accurately but think about what is happening to each side of the shape every time we turn it into a new snowflake.

After each adjustment (we maths nerds call it an 'iteration') how many sides does the Koch Snowflake have?

Okay, something much harder. After each iteration, what is the area of the snowflake and the distance around the outside (its perimeter)?

And, even tougher, but really beautiful — try to understand, as we run this process to infinity, what becomes of the number of sides, area and perimeter of the Koch Snowflake?

Painful penalties

The penalty shoot-out in football is one of the most dramatic, and some people say most unfair, ways to end a sporting contest. Five players from each side take shots at the opposing goalkeeper and soon one side is leaping for joy, while one player on the other team just wishes the world would open up and swallow them.

If you think that's horrible, if the teams are level after 5 shots each, they keep going until one team has scored more goals than the other from the same number of shots. In fact, if all 11 players including goalkeepers have taken a shot, they go back to the start of the list and keep kicking until a decisive advantage is achieved.

The current world record for the longest penalty shoot-out in a first class match saw an incredible 48 penalties taken. It was in the 2005 Namibian Cup when KK Palace eventually beat Civics 17–16. The goalkeepers were on fire that day saving 15 shots between them!

Another procession of penalty pain occurred during the 1988 Argentine Club Championship, when Argeninos Juniors beat Racing Club 20–19 after 44 penalties in the highest scoring shoot-out ever.

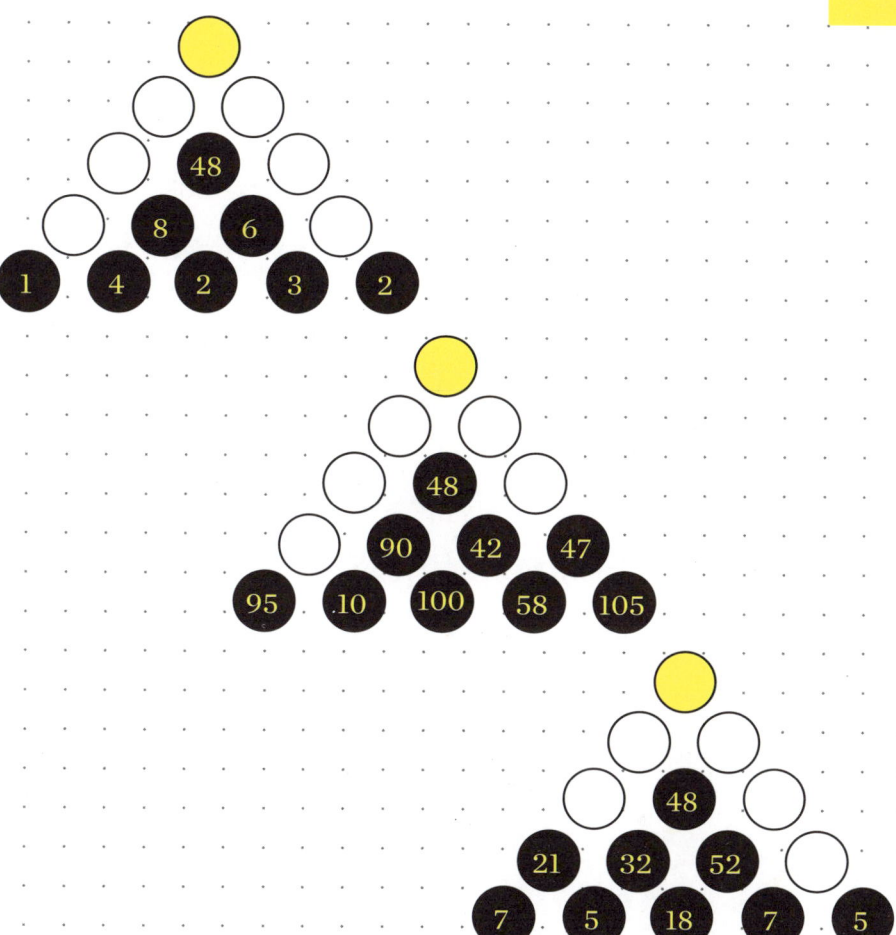

More triangle tangles

This is like our puzzles for 29 ... but on steroids.

In each of the grids above, each circle is filled by a number that is somehow generated by the two numbers below it.

I'm not giving you any more details than that. If this makes no sense to you at all, go back and try the quiz for 29 first. These grids are similar, but the rule is perhaps a bit harder to spot.

Each time you have to crack the relationship between the numbers that would satisfy everything in the grid so far. Then continue this rule until all circles are full.

I want to know, each time, what number goes in the yellow circle?

China produces roughly 49% of the world's apples

… which is about 9 times as much as the next producer, the United States.

If, like me, you freak out at the grocers trying to remember if you're meant to be buying a 'red delicious' or a 'fuji', you'd probably be terrified to know that there are something like 7500 different varieties of apples across the world.

Some of the coolest names for apple types come from classic English apples and include the Beauty of Bath, Greasy Butcher, Cox Orange Pippin, Hog's Snout, Slack-ma-girdle, Peasgood Nonsuch, Broxwood Foxwhelp, Cummy Norman and Fran's Flushed Red Mystery.

Happy birthday, sis

Today is the birthday of sisters Ellie and Olivia.
They have exactly the same birthday, but they are not twins.

Today, Ellie and Olivia turn a combined age of 49.

Ellie is older. In fact, Ellie is now twice as old as Olivia was when Ellie was the age that Olivia is now.

Okay, get up off the floor and read that again: *Ellie is now twice as old as Olivia was when Ellie was the age that Olivia is now.*

How old does Olivia turn today?

*Depp*ends on who you ask ...

Many people know of Johnny Depp. You may know him as Captain Jack Sparrow from the Pirates of the Caribbean *movies, or his superb portrayal of gangster Whitey Bulger in* Black Mass. *If you're my age you might even remember him from the TV show* 21 Jump Street *(yes kids, there was a world before Channing Tatum).*

But one thing Johnny Depp is not, is a qualified bacteriologist. So when on 13 July 2005, appearing as a guest on Jay Leno's Tonight Show *he cited a study done on bar peanuts which revealed the presence of 27 different types of urine, there is a chance he might not have been right on the money.*

Indeed, it seems no such study has ever been carried out. At the same time, hand hygiene is really important, especially after using the bathroom and for staff who handle food. There are many studies that show people are lazy with post-toilet handwashing, or don't do it at all (yuck!). Some studies suggest about 50% of all people don't take a post wee-wee wash.

50%?

Scrub up, people! You're better than that!

A long times

This little maths puzzle involves some good old-fashioned multiplication. Long multiplication.

You remember, the back-to-school, 'put down the three, carry the one' type stuff that we all know and love. Or, if you don't know and love it, I guess you could just grab a calculator. But, hey — where's the fun in *that*?

Seriously, if you do slog these out by hand, it will be very rewarding when you start to see the amazing patterns that emerge.

If you've read any of my previous books you'd be familiar with the Armstrong numbers (which are sometimes also called Pluperfect Digital Invariant numbers ... so you can see why I call them Armstrong numbers).

The number 153 is an Armstrong number because it is 3 digits long and is also the sum of the third powers of its digits. Say what?

Well, 153 has 3 digits, and $1^3 + 5^3 + 3^3 = 1 + 125 + 27 = 153$. Neat, hey? Similarly, $1^4 + 6^4 + 3^4 + 4^4 = 1 + 1296 + 81 + 256 = 1634$, so 1634 is an Armstrong number. You can easily check that 153, 370, 371, 407, 1634, 8208, 9474, 54748, 92727 and 93084 are all Armstrong numbers.

There are only 88 Armstrong numbers, and the largest is the 39-digit brute 115,132,219,018,763,992,565,095,597,973,971,522,401. Yes, $1^{39} + 1^{39} + 5^{39} + ... + 1^{39} = 115,132,219,018,763,992,565,095,597,973,971,522,401$ ... maybe you should just take my word for that.

But let's go back to 153 and have some fun.

Work out (by hand) what $16^3 + 50^3 + 33^3$ equals. Then try $166^3 + 500^3 + 333^3$ and, once you've limbered up, $1666^3 + 5000^3 + 3333^3$ (okay, I'll let you use a calculator for this last one). Pretty cool, hey?

The penny drops ...

Here's one from the United States, where the coins in circulation are affectionately known as pennies (1c), nickels (5c), dimes (10c) and quarters (25c).

I have 50 coins in my purse, totalling exactly $1. I drop one coin. What are the odds it's a penny?

Have a bit of a think, but if you're stuck, here's a hint. You'll need to work out all the combinations of exactly 50 coins that give you a dollar.

Can you have 25c coins in any such combination? What about if 10c coins are the largest in your purse, can you make any combinations of exactly 50 coins? Now try with 5c pieces as the largest, and so on.

Right, has the penny dropped for you?

A devil of a puzzle

The Riddle of the Week on the PopularMechanics.com website is an awesome way to stretch out your grey matter. Jay Bennet posted this classic on 30 March 2017.

Have a go!

One day, while walking home from work, the devil pays you a visit. He has a proposition, and afraid to insult him, you agree to listen.

'I have two 50 dollar bills, and I am going to give both of them to you,' says Lucifer with a sly grin. 'Then I am going to make a statement. If the statement is false, then you give me back just one of the 50 dollar bills. If the statement is true, then you keep them both. Do we have a deal?'

Seeing nothing but profit in the proposal, you agree to the deal, and shake hands with the devil. Satan hands you two crisp 50 dollar bills and makes his statement.

After he does, you owe the devil one million dollars. What did he say?

Hint: the devil's statement gives you a choice between two options, which turns out not to be a choice at all.

Who knew a deal with ol' beelzebub himself would be a bad idea? Find out

Heads, I win!

If a coin is tossed and caught in mid-air it will come up the same way it was facing when it was tossed, not 50% of the time ... but 51%.

This is according to mathematician and magician (genuinely true!) Persi Diaconis and his colleagues Susan Holmes and Richard Montgomery in their research paper 'Dynamical Bias in the Coin Toss'.

If the coin is spun on a table it will land with the heavier side facing down anything up to 80% of the time.

Prime real estate of Area 51

Two mysterious government agents, A and B, are quibbling over the size of their top secret, highly classified, military bases.

They each have rectangular plots with total areas of 51 square kilometres and whole number dimensions, yet Agent A has noticed that the perimeters seem to be different.

Agent B argues that the perimeters *must* be the same, because 51 is *obviously* a prime number.

Agent A takes out his tape measure and proves that the perimeters are in fact different.

So what numbers has Agent A come up with for the perimeter of each of their top secret, highly classified military bases?

Australia *no longer* has the number 51 fastest internet speed in the world …

We've shot up to 50!

The final Akamai State of the Internet report for 2016 made international headlines when it showcased that despite the National Broadband Network being rolled out for several years now, Australia had the 51st best internet speed in the world.

Well, in the first quarter of 2017 we had actually edged up to 50th, but still sit well behind Estonia (46th), Kenya (43rd), Malta (38th), New Zealand (27th), Thailand (21st), Latvia (17th), Romania (13th) and cannot even peer far enough into the distance to see the perennial leaders, South Korea.

Just Cos

Even the oldies reading this book should be able to remember from high school the basics of trigonometry, our good friends sine, cos and tan. They tell us things about the relationship between angles and the shapes in which you may find them.

You may even be able, deep within the cobwebs, to remember drawing a right-angled triangle with two sides of length 1, or a side of length 1 and a side of length 2, and applying a bit of good old Pythagoras' theorem. Using cos to refer to the ratio of adjacent side of the triangle to the hypotenuse, we get:

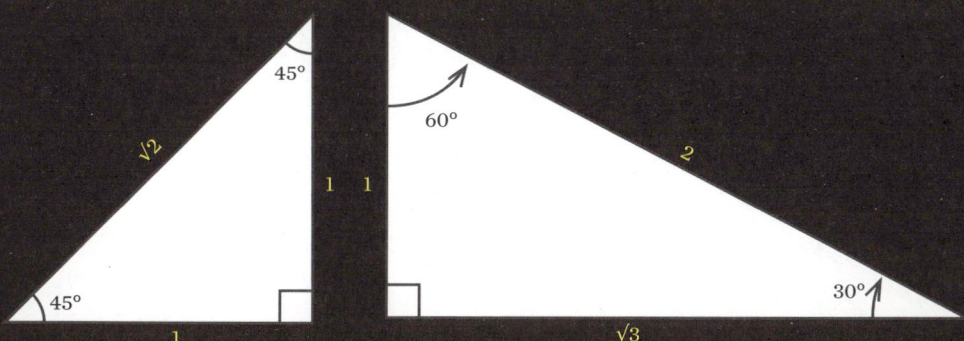

cos 30° = $^{\sqrt{3}}/_2$, cos 60° = $^1/_2$ and cos 45° = $^1/_{\sqrt{2}}$

6 fer 52?

For 3 magic days from 23–25 February 2017, the cricket world held its breath as Australia did the unthinkable and smashed India, in India, by a whopping 333 runs, requiring only 3 of the 5 days allocated for the match.

There were many amazing performances by the Aussies, not the least captain Steve Smith's second innings century wherein his score of 109 eclipsed that of the entire Indian team in each of their innings (105 and 107).

But the man of the match was the previously unheralded New South Wales left-arm spinner Steve O'Keefe who tore through the Indians in the first innings in one glorious 23-ball spell taking 6 wickets for just 5 runs on his way to a haul of 6-35. Come the second innings the man they call SOK starred again, nabbing another bag of 6-35.

As a cricket-mad mathematician I was blown away by the massive Australian win, but also stunned by the symmetry of O'Keefe claiming the exact same figures 6-35 in each innings. It got me wondering if these are the best ever 'repeated figures' in a test match.

'But who could you ask such an obscurely nerdy cricket question, Adam?' I hear you cry. Well, Ric Finlay of the ABC, Australia's pre-eminent cricket statistician, of course. A friend of a friend got me Ric's number and I texted him my query. Within 5 minutes I had the response which I will reproduce here in full:

'G'day Adam. Yes SOK has the best repeated figures in tests now. Previously held by Bhagwat Chandrasekhar, 6-52 Ind v Aus, 1977–78 MCG. He took the mantle from Pakistan's Intikhab Alam, 5-91 v NZ, Dhaka, 1969–70. Ric'

I told you Finlay was the guru of cricket stats.

Well played, SOK. And love your work, Ric.

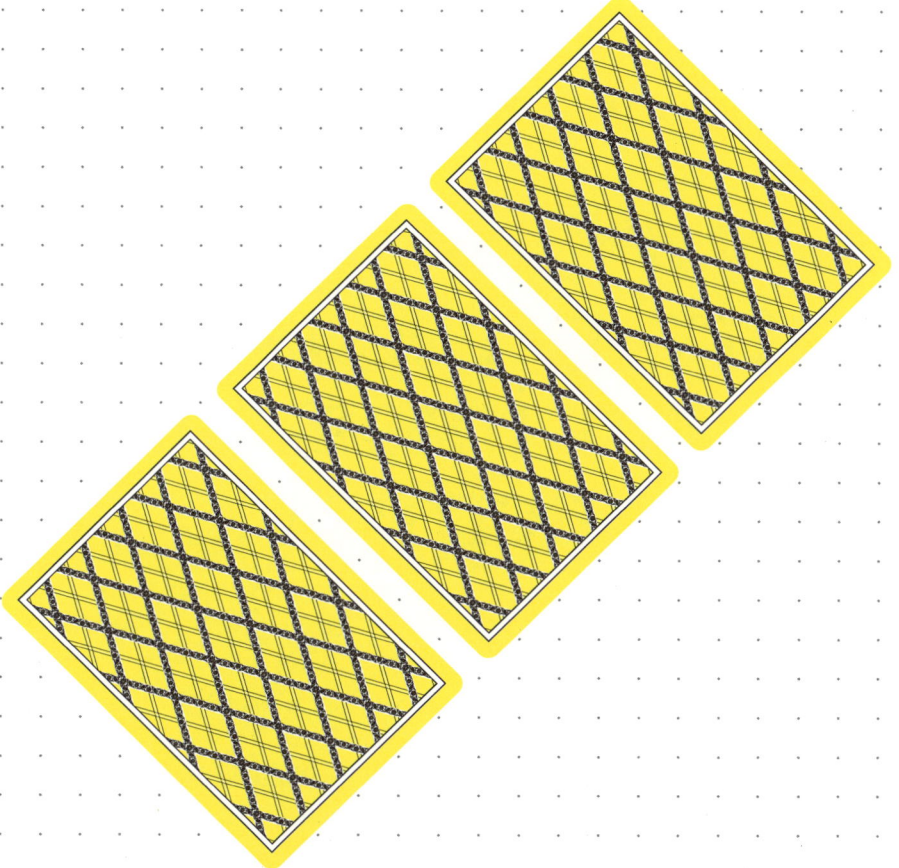

Play your cards right

There are 3 playing cards lying face up, side by side. A 9 is immediately to the right of a 3. A 5 is not next to a 9. A spade is immediately to the left of a diamond, and a spade is immediately to the right of a spade.

If I add up the value of all the spades, what number do I get?

While we're playing with cards, you know how a regular deck has 52 in it? Tell me, how many ways can you select 3 cards from it?

Fifty-three is its mirror image in hexadecimal ... *huh*?

The fact that we count in base 10 is so essential to our mathematics that most of the time we don't even realise that we are doing it. Base 10 means that after the 10 digits 0, 1, 2 ... , 8 and 9 we are finished with all the 'units' and we move on to the tens. Once our tens are full up to the 90s we hit 99 then we are finished with the tens and we move onto the hundreds. And so on.

This means that when we write a number like 4381 we are representing 4 × 10 × 10 × 10 + 3 × 10 × 10 + 8 × 10 + 1.

Well, let's imagine we counted in base 16, which is called hexadecimal and arises in computing because 16 is a power of 2. We would need extra symbols in our digits before we filled them up and it is traditional to use A, B, C, D, E and F for what we base 10ers would consider the numbers 10 to 15.

So the number 53, when written in terms of its powers of 16 is 3 × 16 + 5. This, in hexadecimal, is written as 35.

So yes, 53 is its mirror image in hexadecimal!

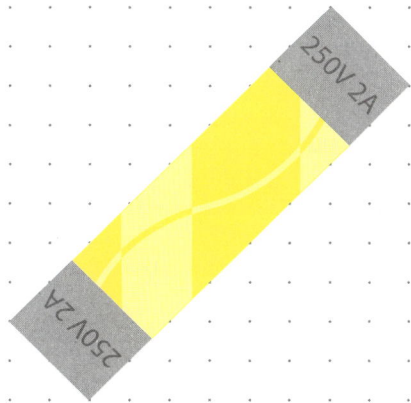

En*fuse*iastic Ellie

Ellie the electrician got to the plant as quickly as she could. The manager was very glad to see her.

'Thank goodness you're here! It's a disaster! There are fifty-three fuses on our power board and over half of them have blown — they just aren't responding at all. Ten of the first 14 are busted. But be careful! Only remove a broken fuse. If you touch a live one, you'll get fried!'

Without hearing anything further from the manager, Ellie calmly took out one of the fuses and began to examine it.

How did she know which number fuse to remove without getting fried?

Where's there's a Liouville, there's a way!

Here is a beautiful little mathematical oddity discovered by the great French mathman Joseph Liouville.

Liouville made many discoveries in a wide range of mathematics, established one of the world's most famous mathematical journals and even has a crater named after him on the Moon.

He also came up with this little bit of fun.

Write down any number and under that write down its divisors: for example, if I chose the number 6 which has divisors 1, 2, 3 and 6, I'd write:

6
1 2 3 6

Now write under this the number of divisors that each of these divisors has.

This sounds ugly but 1 is its only divisor, 2 and 3 are prime so have only two divisors being 1 and themselves and we've already worked out that 6 has 4 divisors. So now we have:

6
1 2 3 6
1 2 2 4

Now show that the square of the sum of this last group of numbers equals the sum of the cubes of these numbers. Huh? What it means in this case is that:

$(1 + 2 + 3 + 4)^2 = 100$ and $1^3 + 2^3 + 3^3 + 4^3 = 1 + 8 + 27 + 64 = 100$.

Neat, hey? Here's a slightly more complicated example for you, using the number 12.

Here we get the 3 rows:

12
1 2 3 4 6 12
1 2 2 3 4 6

And you can easily calculate that:

$(1 + 2 + 2 + 3 + 4 + 6)^2 = 1^3 + 2^3 + 2^3 + 3^3 + 4^3 + 6^3 = 324$.

I'm pretty sure you can see what's coming.

Show that this process holds up in the more difficult example of ... yep, good old number 54.

Now, 54 has a lot more divisors than 6 or 12, so please be thorough. Good luck!

And remember: you know what they say ... where there's a Liouville, there's a way!

Thomas (died far too) Young

On 10 May 1829, the incredible Thomas Young died, aged just 55.

This brief entry will not do justice to the mind of Thomas Young, who was called by one author 'The Last Man Who Knew Everything'.

Here are just two achievements by the great T-Yo.

In 1814 he translated one of the 3 scripts on the famous Rosetta Stone. The stone had been recovered in 1799 and had the same information written in 3 languages. Because of the work of Young and others in translating all 3 scripts, we gained invaluable insight into the hiero-glyphic language of the ancient Egyptians.

Now that's pretty awesome. But even more amazing IMHO was that 10 years earlier, Young had been one of the key players in establishing the 'wave theory of light'. In a famous 'double slit experiment', he showed that when you shine light through one or two very narrow grooves in a piece of cardboard and onto a wall, the resulting pattern of light on the wall has brighter and darker segments. This shows that the light is trav-elling in waves which cancel each other out at some points and amplify brighter in others.

Up until this point light had been thought to travel only as a series of particles. The double slit experiment is so amazing that the great physi-cist Richard Feynman said if you think about it hard enough you can pretty much understand all of quantum mechanics!

Nice work, Youngster! And I haven't even touched on your work in how we perceive colour, what makes things elastic, equations for surface ten-sion, the difference in medicine dosage sizes for adults and children, how the human pulse beat works, how to tune musical instruments ... okay, slow down, Thomas. No one likes an overachiever!

A family affair

What a lovely day this is.

Dad is sitting down with his twin daughters to celebrate their birthdays. That's right, they all share the same birthday, which makes birthdays even more special in this house.

But Dad, ever the loveable maths geek, can't help turning the day into an equation.

'Well, girls,' he says, 'today we bring up 77 years between us. But it feels like you were born only yesterday. Of course you weren't. But do note that on the day you were born, I was turning twice the age that you are now combining to be.'

The twins pretend to be paying attention, but honestly, they are just throwing down birthday cake.

Help me work this out. How old are the twins turning?

1	2	3	4	5	6	7	3
5	8	7	8	1	1	4	5
6	4	6	8	6	2	2	8
3	7	2	5	3	7	8	6
3	7	2	1	4	3	2	1
6	8	4	4	6	5	5	7
5	1	1	3	7	8	2	4

Jigsaw puzzle

Everyone learns back in school, when you first wrestle with your times tables, that 7 × 8 = 56.

The 56-square grid above is made up of 7 pieces of 8 blocks each. Just in case you're wondering, such blocks of 8 squares are called 'octominoes'.

On the opposite page, we have 7 different octominoes, (two are just rotations of each other) selected from the 369 different octominoes.

Now, as they lay across the grid above, each block contains the numbers 1 to 8, exactly once each. The numbers don't have to follow any logical order within each piece and the pieces may be rotated or reflected from how you see them here.

Show how the 8 blocks fit together across the grid.

56-gon typhon

*The shape with 3 regular (or equal) sides is the equilateral triangle.
Similarly, we say the 'regular 4-gon' is the square. The 50 cent piece
in your pocket is a 'regular 12-gon,' or 'dodecagon'.*

*Because of the factorisation of 68 as $2^2 \times (2^{2^2} + 1)$, a 68-sided regular
polygon may be constructed with compass and straightedge. Though
to properly explain why and how would probably take a whole book's
worth of hardcore maths. Sadly, the 56-gon is not constructible with
compass and straightedge.*

*According to the famous historian Plutarch, the equally famous
mathematicians known as the Pythagoreans associated a regular 56-gon
with the mythical creature Typhon — a monstrous giant with snakes on
his shoulders, described variously as 'terrible, outrageous and lawless'
and the most deadly creature in the whole of Greek mythology.*

*The connection between Typhon and a 56-sided figure is lost on me,
but given the rap sheet Typhon carries, if he wants the 56-gon,
I ain't standing in his way.*

*A famous 56-gon of sorts exists at the 5000-year-old Aubrey Hole Circuit
in Stonehenge. The Circuit consists of 56 chalk pits laid out in a very
accurate circle and some people think it assisted Stonehenge operating as
a 'neolithic computer' with which eclipses could be calculated and the like.*

*I'll stop now, because I'm afraid I might be getting a little
bit too close to the truth.*

Piece of pizza?

Summer has long gone, the leaves are falling from the trees and our galloping gourmands gather again for their autumnal ordering.

The location is Piettro's Puzzle Place Pizza Palace located at 57 Puzzle Place. Again the 8 people tucking into the garlic bread and bruschetta around the table, in no particular order, are Aaron, Barry, Caroline, Danielle, Eamon, Fantine, Gary, and Hayley.

I can tell you that:

- Aaron sat in seat number 1 opposite Hayley.
- The 4 couples happen to all be male — female. Their couplings are, with the female named first (Caroline/Aaron, Fantine/Barry, Danielle/ Eamon and Hayley/Gary). On this occasion no two couples sat the same distance apart. Hint: think about what this crucial clue must mean.
- Danielle sat closer to Barry than she sat to Gary.
- Neither Danielle nor Hayley sat next to men.
- From Aaron's perspective, Danielle was on the left half of the table.

Can you tell me the seating arrangement?

When is a prime number *not* a prime number?

When it's Grothendieck's Prime!

Alexander Grothendieck was perhaps the greatest mathematician of the 20th century. An all out, bona fide, superstar genius who created massive amounts of new mathematics that are central to the science today. But that doesn't mean he couldn't make the occasional mistake.

Grothendieck thought about mathematics in a very abstract way and a lot of it had very little to do with things as everyday as counting numbers. According to one story, a colleague was once discussing a mathematical theorem with him and suggested as an example they should consider a prime number.

'You mean an actual number?' Grothendieck asked and when his colleague replied 'yes, an actual prime number,' Grothendieck said, 'All right, take 57.'

57 of course is not prime. It is 19 × 3. But today, 57 is known among mathematicians as 'Grothendieck's prime'.

Careful what you wish for ...

58 was the age at which American Independence-seeking politician James Otis Jr. wrote in a letter to his sister Mercy Otis Warren in April 1783, 'My dear sister, I hope, when God Almighty in his righteous providence shall take me out of time into eternity, that it will be by a flash of lightning.'

58 was also the age of American Independence-seeking politician James Otis Jr. when, while standing out the front of a friend's house on 23 May 1783, he was killed by a bolt of lightning.

Puzzle like a MANIAC!

Okay, remember the barrel of fun we had back for the number 45, when we tried to fit the 12 free pentominoes into various grids?

Well, those rectangular grids aren't the only awesome shape that can be made.

In 1958, guru of Mathematical Logic, Dana Scott, using the state of the art MANIAC computer (a whole 64K of memory, baby, but a mere fraction of the size of ENIAC, its older sibling pictured above), showed exactly how many different ways the 12 free pentominoes can fill an 8 × 8 grid with the middle 4 squares removed.

They can do this 65 different ways (ignoring rotations and reflections) but there are only 3 possible places that the 'X' pentomino can go.

Try and find the 12 free pentominoes in these 3 grids and, when you're done, you'll have one example of each of the possible locations of the X pentomino.

To help you, each pentomino contains the letters 'OMINO' in some arrangement. Your time starts ... now!

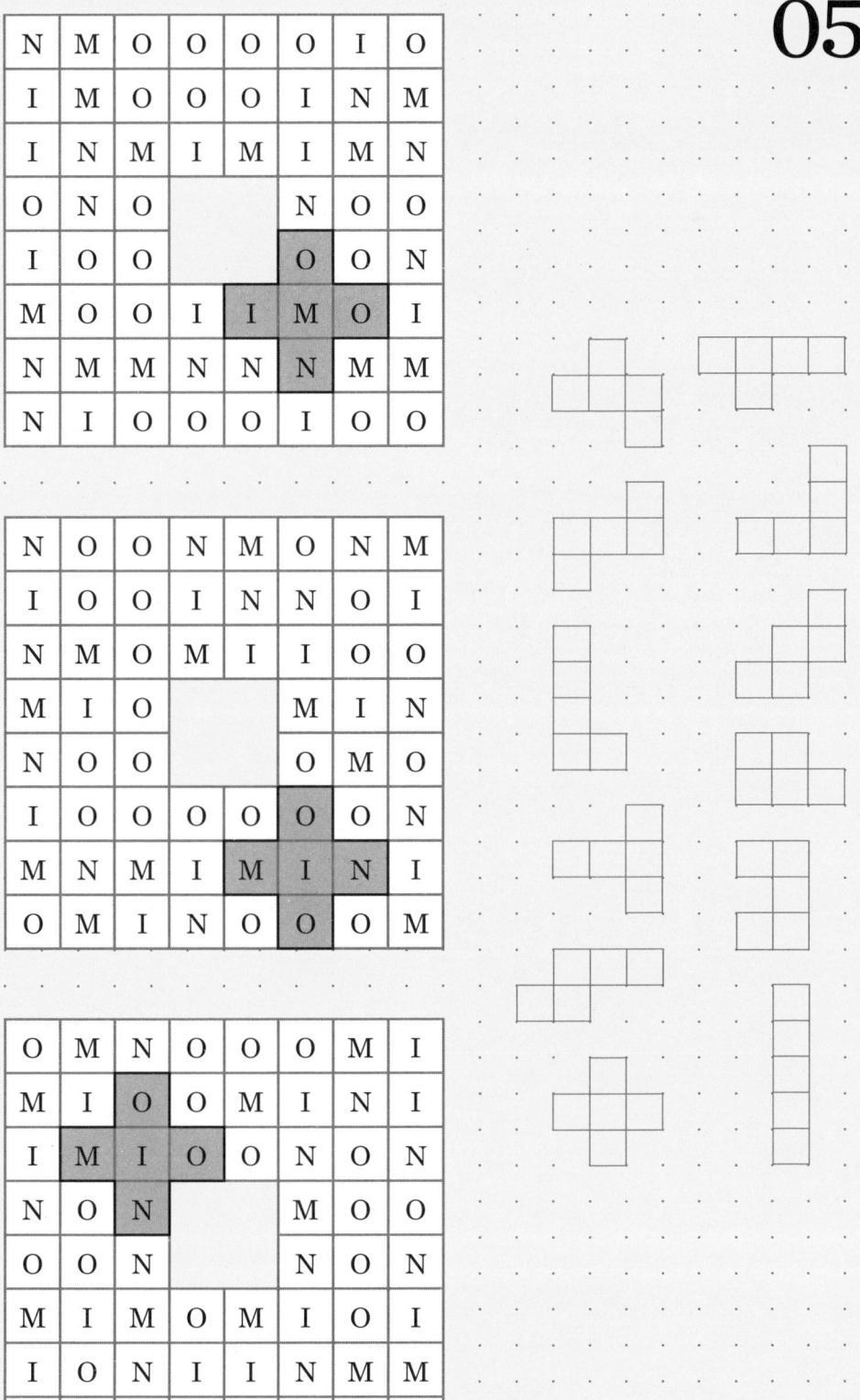

The Sydney Harbour Bridge is officially listed as being 49 metres wide, 1149 metres long and weighs 52,800 tonnes

The massive steel bars that run from the top of the arch to the road level of the bridge are called hangars and the are each 18.28 metres apart dividing each arch into 28 panels. The longest hangar is just under 59 metres, the shortest under 8 metres.

But here's the cool, or hot, thing about the bridge. While its height is officially listed as 134 metres above sea level, in extremes of temperature, the top of the arch can move up or down by as much as 18 centimetres.

$$\underline{} \times \underline{} + \underline{} + \underline{} + \underline{} + \underline{} - \underline{} = 59$$

$$\underline{} \times \underline{} \times \underline{} \div \underline{} + \underline{} \times \underline{} - \underline{} = 59$$

$$\underline{} \times \underline{} - \underline{} \times \underline{} + \underline{} \times (\underline{} + \underline{}) = 59$$

$$(\underline{} + \underline{}) \times \underline{} - (\underline{} + \underline{}) \times (\underline{} - \underline{}) = 59$$

$$(\underline{} \times \underline{} \times \underline{} + \underline{} + \underline{} - \underline{}) \div \underline{} = 59$$

They're baaaack

You might remember these digital equations from waaaay back at number 32.

Well, here are some more. But, be warned ... these are a little harder than before. You need to place 1, 2, 3, 4, 6, 7 and 8 once each into the equations to get the answer 59. Limber up and grab your pencil!

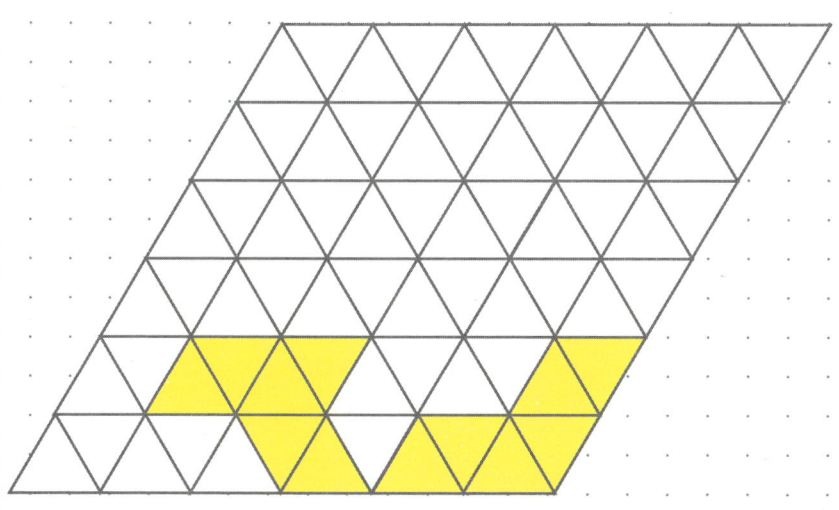

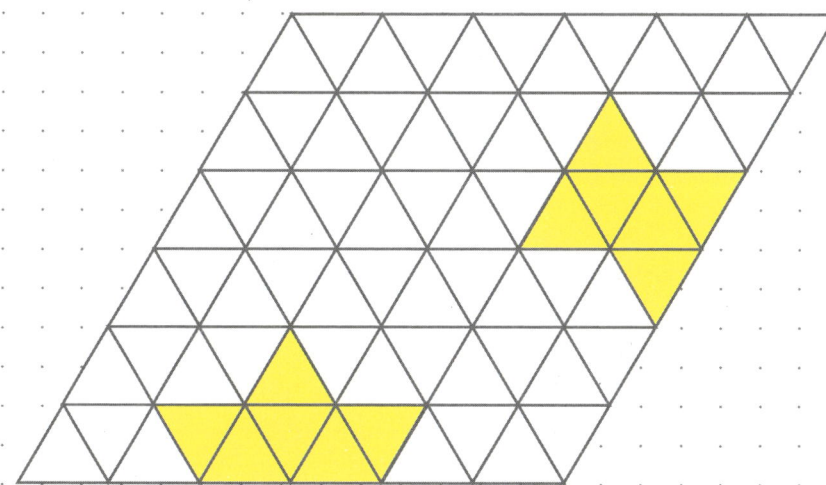

Hexiamonds are forever

What's a hexiamond? Glad you asked.

A hexiamond is a polyiamond composed of 6 equilateral triangles. And a polyiamond? It's a polyomino comprised of equilateral triangles. What's a polyomino? Hey, if you're reading this book in order, you should already know this one. But if not, flip to 005 and come back ready to tackle the beauties above. Your job is simple: fit the pieces into the grids.

In these grids, we've provided the position of two tiles. See if you can fill in the remaining 60 spaces with the 10 unused hexiamonds.

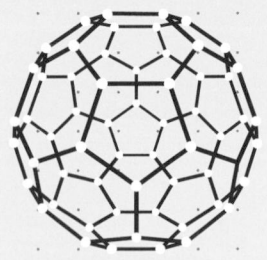

And the hardest, stiffest and densest substance known to humanity is ...?

There's a good chance you said 'diamonds', and while you're not strictly correct, you're pretty close.

Diamonds are created when carbon, in the form of graphite, is subjected to intense heat and pressure. The atoms of carbon in a diamond are arranged in cubes stacked in a pyramid.

But hyperdiamonds are created by applying incredible heat (over 2000 degrees Celsius) and pressure (200,000 times that in normal atmosphere) to a form of carbon called fullerite.

Named after the geodesic domes of architect Buckminster Fuller, fullerite sees carbon arranged in 'buckyballs' of 60 atoms at a time and in hyperdiamonds these buckyballs form tiny interlocking 'nanorods'.

One last thing to know about diamonds, while jewellers like to convince would- be customers that 'diamonds are forever', in fact, all diamonds are — incredibly slowly over very long periods of time — turning back into graphite.

Big Easy fun

No book by me would be complete without mention of my maths-crush, AKA the greatest maths guru of all time — Leonhard Euler.

Way back in 1754, the Big Easy (no, he didn't call himself that — he went by more impressive titles like 'the master of us all') proved that 'every prime of the form $4n+1$ is the sum of two squares'.

Don't be scared of this.

It just means that a prime like 61, which can be written as $4 \times 15 + 1$ can also be written as the sum of two squares. In this case, $61 = 5^2 + 6^2 = 25 + 36$.

We call these sort of prime numbers 'Pythagorean Primes'. Another brilliant maths mind, Johann Peter Gustav Lejeune Dirichlet (who is just known as Dirichlet these days, which is a frightful waste of some awesome names), showed that there are infinitely many Pythagorean Primes.

Your challenge is to find all of the Pythagorean Prime numbers from 1 to 100. For each of them, write the prime as the sum of two squares.

Sometimes ...

when your phone or computer crashes, it may actually be because of some of the 61 types of subatomic particles flying at you from outer space!

In 1932, James Chadwick bombarded atoms of the element beryllium with special particles known as alpha particles. In doing this, he discovered the neutron and so in 1932 we knew of 3 basic particles that were smaller than atoms (or subatomic). They were the electron, the proton and Chadster's little find, the neutron.

Fast forward to 2013 and the discovery of the Higgs boson. In the space of just one human lifetime we have gone from 3 subatomic particles, to 61. I wrote about these in my book World of Numbers *('but where can I get that, Adam?' ... 'At adamspencer.com.au, dear reader'). It can get a bit heavy trying to understand fermions and bosons but pause for a second to realise that going from 3 subatomic particles to 61 is a beautiful crystalisation of the amazing time in which we live.*

Actually, they aren't all just fun and games. It turns out these gorgeous gremlins could also crash your computer. At the 2017 meeting of the American Association for the Advancement of Science in Boston, Doctor Bharat Bhuva presented 'Cloudy with a Chance of Solar Flares: Quantifying the Risk of Space Weather'. According to Dr Bhuva, cosmic rays which fly across the cosmos (from the hearts of supernovae among other things) and hit the Earth's atmosphere, give off showers of other subatomic particles. A small fraction of these carry enough energy to affect the circuits in our digital devices where they could change a 0 to a 1 and even make your system crash.

It's probably not a common cause of system failure, but next time your screen freezes up except for the spinning pizza wheel of death, simply shout, 'damn you, cosmic rays'. Won't help you save your files, but you will look pretty smart.

5	×	2	+	3
−	2	−	8	×
9	×	6	−	3
+	1	×	2	=
5	+	2	=	62

6	−	4	−	8
×	8	×	5	−
3	×	2	−	5
−	2	×	1	=
9	×	8	=	62

2	+	4	×	7
+	5	−	1	−
4	×	5	−	9
×	3	+	8	=
3	×	7	=	62

Sixty-two snakes

Cast your mind back to a distant time ...

Okay, maybe just as far back as 28. Remember these slippery snakes?

Same deal here: you can move up, down, left or right, but you can never visit the same square twice.

See if you can find the 3 different paths that each trace out the correct equation.

Geek disclaimer: order of operations do apply. Slither on!

Here's a spooky fact about the number 62 ...

it is the only number whose cube (62³ = 238328) consists only of 3 digits each occurring exactly 2 times.

Another spooky fact often claimed about 62 is that the famous 'father of psychoanalysis' Sigmund Freud, was scared of it. In a letter to his colleague Carl Jung, just after Freud had been assigned a phone number ending in 62, he wrote 'it was plausible to suppose that the other figures signified the end of my life, hence 61 or 62. Suddenly method entered into my madness. The superstitious notion that I would die between the ages of 61 and 62 proves to coincide with the conviction that with "The Interpretation of Dreams" I had completed my life work, that there was nothing more for me to do and that I might just as well lie down and die.'

But before you get too excited about this it should be noted that for a while there, the Sigster was also down on the number 67.

And, one final reality check, if someone regurgitates to you the urban myth that 'Freud feared 67 ... and died at exactly that age', feel free to point out that he actually lived until the age of 83.

After-school *advenn*tures

All 63 students in year 7 at Puzzler's High are offered after school activities for second term.

There is football on Tuesday, music on Wednesday and chess on Thursday.

Thirty-three students sign up to play football, 34 sign up to do music and 37 want to join the chess club.

But some of these students are doing more than one activity. Fourteen are doing football and music, 16 are doing football and chess and 16 are playing music and chess. In fact, 4 of these eager-beaver students are doing all 3 activities.

Are any students doing no after school activities? If so, how many?

To qualify as a cyclone ...

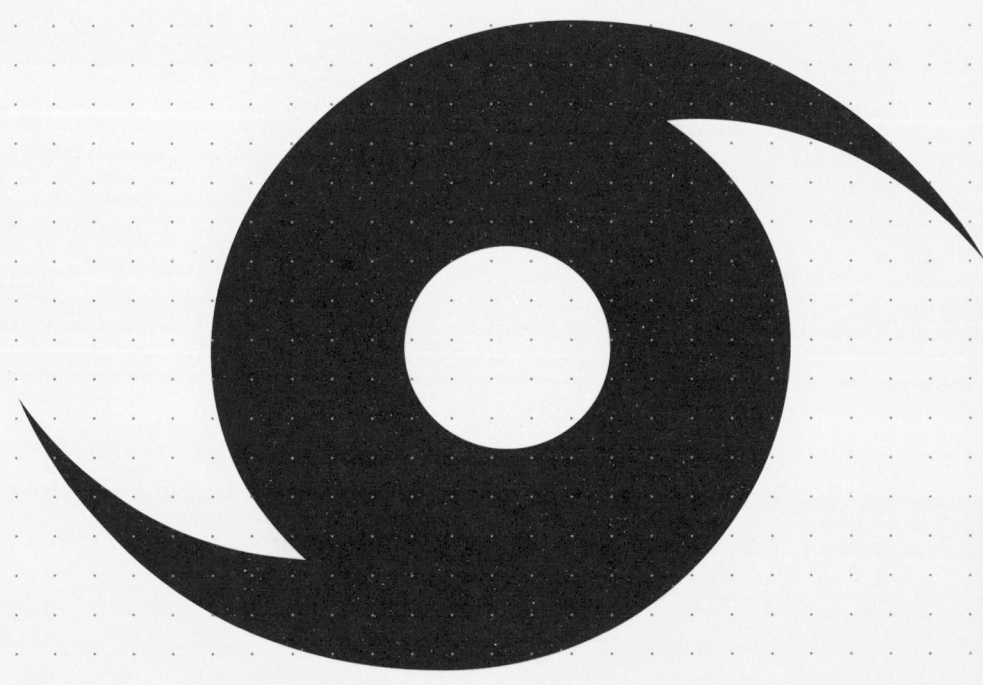

... you need a low pressure system to form over warm tropical water accompanied by gale force winds (sustained winds of 63 kilometres per hour or greater), and gusts in excess of 90 kilometres per hour near the centre.

• 21	• 21	• 21	• 10	• 11
• 10	• 11	• 3	• 9	• 9
• 15	• 16	• 20	• 40	• 23
• 21	• 11	• 40	• 23	• 12
• 21	• 21	• 21	• 11	• 17

Cold as splice

Remember this little grid from back at 26?
Your job is to split the array of numbers above into sections so that each non-empty region's numbers sum to, no, not 26. Yep, now they have to add up to our magic number 63.

To mark off these regions you may only use 3 lines. Each line must be straight and run between two yellow dots. And if a line goes through a black dot, that dot's number has been crossed out and doesn't belong to any region.

It's the dots that score the points. If a dot is in a region it keeps its number even if the number is crossed out or lies in another region.

Off you go!

8^2

Sixty-four was the number worn by the NFL's Baltimore Ravens guard John Urschel

He's 191 centimetres tall, weighs 136 kilograms and, in a rare moment of science trumping sport, recently retired from the game to focus on completing his PhD in Applied Mathematics at MIT, two days after a damning medical report found that 99% of NFL player's brains studied had tested positive for the degenerative brain injury, chronic traumatic encephalopathy, or CTE.

He's also a handy chess player, loves numerical linear algebra and can bench press 100 kilograms 30 times. RESPECT.

YouCube

If you've stopped by number 27, you'll know I don't mind a spot of cubist painting.

Well, now I have a cube of side length 40 centimetres. I paint all 6 faces of the cube yellow (I'm in a yellow period). I then cut along each side every 10 centimetres to produce 64 smaller cubes of side length 10 centimetres.

I throw all the smaller cubes into a sack, shake them about, close my eyes and pull out a cube. What is the most likely number of yellow faces that cube will have?

When Apple updated to the iOS11 operating system, over 100,000 apps died a digital death

iPhones have run on 64-bit chips since the iPhone 5s used an Apple A7 chip in 2013, but the new iOS11 operating system only allows apps that run on 64-bit processors. This is why iPhone users had been receiving messages for a while saying 'The developer of this app needs to update it to improve compatibility. This app will not work with future versions of iOS.'

So it's time to say goodbye to Flappy Bird, True Skate, and Ridiculous Fishing. True friends you were, but time has passed you by. It's also probably time to put that awesome looking iPhone 5 you've been rocking ... back in the drawer.

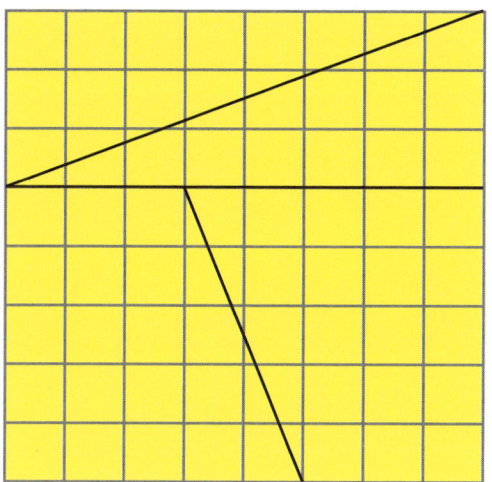

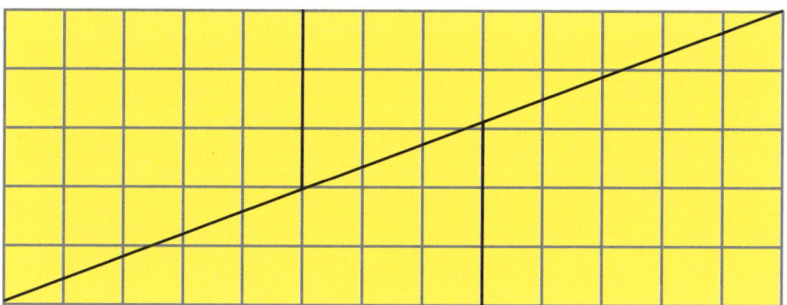

Use your illusion

In this famous maths problem I've taken an 8 × 8 square and cut it into 4 shapes — two equal triangles and two equal trapeziums (or you can say trapezia ... whichever's ezia to say! Ha! Get it? Trapezia ... ezia ... easier! Boom, yeah!)

I've then taken those 4 shapes and rearranged them into a rectangle of area 5 × 13.

But the original square had area 8 × 8 = 64 square units. The rectangle measures 5 × 13 = 65 square units.

How can these four shapes make up a square and a rectangle which are different sizes? How can 64 = 65?

A mature Crown of Thorns Starfish can produce up to 65 million eggs in a single season

One of the many threats to the natural beauty that is the Great Barrier Reef is regular outbreaks of the Crown of Thorns Starfish. Indeed, it may be responsible for up to half of the coral loss experienced by the reef over the last 30 years.

It's complicated because the Starfish can also play a positive role. It tends to eat the faster-growing coral types, allowing the slower ones to grow and thus increase the diversity of coral types on a reef.

But too much of anything, especially Crown of Thorns Starfish, can be a bad thing and it seems around every 14 to 17 years there are outbreaks which see the reef marauded. Outbreaks once would have occurred every 100 years or so, but human influences, like increased levels of nitrogen in the water from agricultural run-off, have driven Starfish numbers up to perhaps a whopping 60 million currently feasting on the reef.

This along with the warmer waters driven by climate change have seen the Great Barrier Reef take a battering of late. We can only hope that we change our behaviour ... and quickly ... to save this natural wonder.

7	1	7	3	7	6	6	5
1	3	2	6	7	6	1	2
5	2	1	6	1	7	6	5
6	6	5	1	1	7	3	5
4	4	4	2	5	7	7	1
3	1	6	6	6	3	5	3
3	7	1	7	5	2	1	1
1	7	6	4	1	3	6	1

6	7	3	1	6	6	2	2
1	7	6	4	7	1	2	3
6	5	4	1	1	5	3	3
1	4	4	3	6	6	7	4
7	2	5	1	4	5	7	6
6	2	2	6	3	3	7	6
1	1	3	4	5	1	3	5
5	5	4	7	1	4	5	7

Divide to conquer

How'd you go with those divisions at number 42?
Time to sharpen your pencil and divide these little babies up.

Your goal is the same — divide each grid into 4 identical shapes of 16 squares each. No prize for guessing that the sum of their interior numbers for these should equal our new target number — 65. And if you thought the shapes were hard to find at 42 ... brace yourself.

Go ahead and divide to conquer!

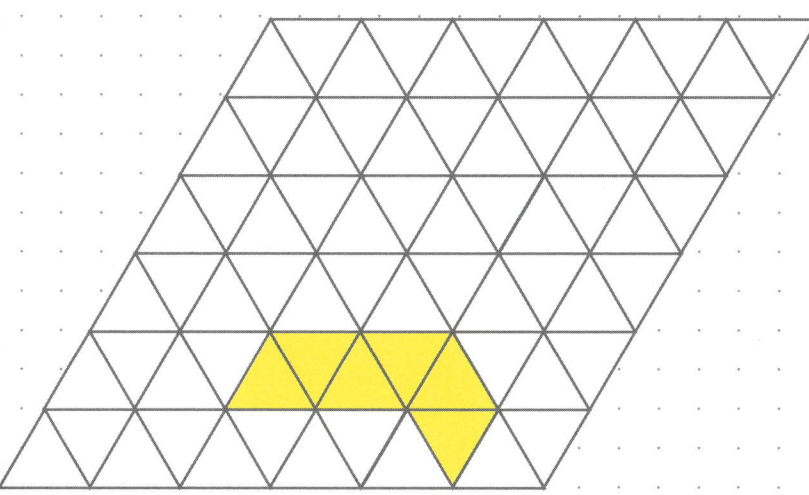

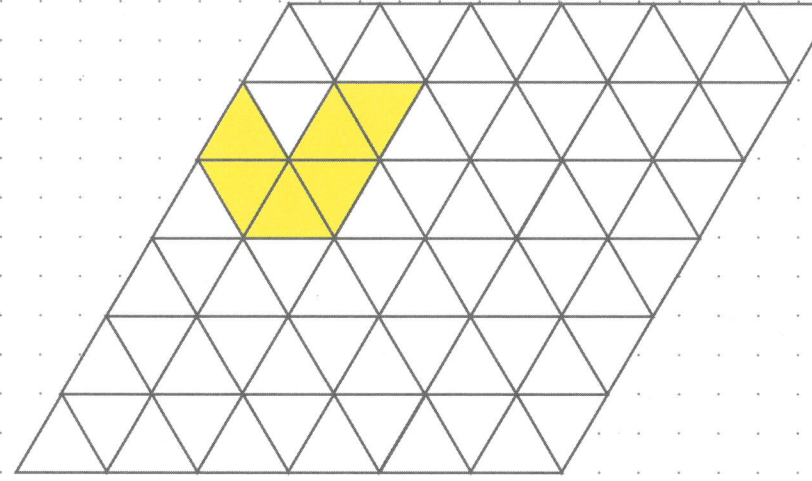

Hexiamonds are a geek's best friend

You might have guessed back at number 60 that there are many ways to tile certain grids with the 12 different hexiamonds.

In the grids above, we have once again provided the position of some tiles. Go ahead and fill in the remaining *66* spaces with the 11 unused hexiamonds.

Hey, and if you'd like to download a copy and go crazy with your scissors, you'll find PDFs of the pieces on my website.

Welcome back!

Here we are, back again with a few more jigsaw puzzles. If you haven't already tackled the ones at 29, go and give them a shot first, then come back. I'll wait ...

Welcome back! Your task is the same. Try to fit the yellow jigsaw pieces into each of the 3 puzzles. The rules remain the same: match each number to its twin. You can also rotate the pieces to create a match, but there are no reflections.

Once again, I've started you off with the one below to ease you into things. Once again, there are no rotations at all in the first one. But you can bet your bottom dollar, there are in the other two!

Hit it!

066

			3		
3					3
			1		
			3		

					2
					3
2					

3	1
2	3

2	1
2	1

3	2
1	2

1	2
1	1

2	3
1	3

2	3
2	1

1	3
2	1

1	1
3	2

2	1
3	1

Lance 'Buddy' Franklin wore guernsey number 67 in the 2017 Sir Doug Nicholls Indigenous Round

He wore it as a special nod to the 1967 referendum that voted overwhelmingly to include Aboriginal Australians in the census and allow the Commonwealth Government to make laws concerning Aboriginal people.

Buddy normally wears jumper number 23 but along with 9 other players from various clubs changed to 67 for the night. Teammate Dane Rampe was one of several players across the league who swapped his 24 for a 50 to symbolise the 50 years since the historic vote.

The $20K question ...

Remember the prestigious Mathcounts National Competition speed maths battle to the death we came across earlier?

Well, on the off-chance you found the last puzzle too easy, try the final brain buster from Mathcounts 2014:

The smallest integer of a set of consecutive integers is –32. If the sum of these integers is 67, how many integers are in the set?

I don't expect you to smash out the answer in 12 seconds, like 13-year-old Swapnil Garg did to win Mathcounts 2014 and pick up a handy $20,000 scholarship. But if you take some time to think about this you can crack it.

Have a go, but if you're really stuck, here's a hint. Think about the long line of integers starting at –32 then –31, –30, ... up through 0 and into the positives. As you add them together what happens when the positive numbers start being involved in the sum. Good luck.

Concatenation ... continued

Back at 34 we met a couple of cute concepts, namely concatenation (joining numbers like 3 and 4 to give us 34) and summing the numbers 1 through 9 in numerical order to give us 100.

Hopefully you used the hints I gave to get the equation:

$$1 + 2 + 34 - 5 + 67 - 8 + 9 = 100.$$

It turns out our buddy 67 occurs a lot in the 12 equations that satisfy the above conditions.

Apart from the equation I've just given you, can you find the other 4 equations involving 67?

Don't forget the other powerful observation we made in getting the above answer. This saves you heaps of time when you're trying to simplify what you need to crunch out here. When you take the sum of a group of numbers, then turning any + signs to negatives or – signs to pluses will change the sum but not affect whether the sum is odd or even. Read that again. So:

$1 + 23 + 4 + 5 + 8 + 9 = 50$ which is even. So any changes to the + and – signs will still be an even number. These numbers 1, 23, 4, 5, 8 and 9 cannot give us the 33 we need to add to 67. Not in any combination. You don't even need to consider it.

But $1 + 2 + 34 + 5 + 8 + 9 = 59$ which is odd so perhaps can be adjusted to give us the 33 we need.

Also as a further hint, you can concatenate 3 numbers together. In fact, 3 of the 4 solutions you'll see involve a 3-digit concatenation. In fact, one of them involves 4 concatenated numbers.

Off you go!

68	51	63	68	47	63
63	47	47	51	51	68
63	68	68	51	51	47
47	51	63	47	68	63
47	63	51	47		
51	68	68	63		

47 51 63 68 ... huh?

Check out the grid above.
Which of the remaining squares, if any, should contain a 68 if the pattern is completed?

A spot of recreational maths ...

One of the most famous pieces of unsolved 'recreational mathematics' is a claim by the 18th century German mathematician Christian Goldbach. In June of 1742 he was Snapchatting with the greatest mathematician of all time Leonhard Euler (okay, they weren't using snapchat, they were exchanging letters, but you get my drift) when Goldbach suggested that 'every even integer greater than 2 can be written as the sum of two primes'.

Euler agreed that this looked to be true, but said he couldn't prove it. Well, in the 275-plus years since, no one else has either.

While 4 = 2 + 2 and 6 = 3 + 3 and 8 = 3 + 5 and 10 = 3 + 7 = 5 + 5 and so on and so on, no one can show that every even number greater than 2 can be written as the sum of two primes. We have shown it for even numbers up to 4,000,000,000,000,000,000 but that doesn't mean it happens for all even numbers!

If you had managed to solve this between 20 March 2000 and 20 March 2002, Faber and Faber would have paid you a $1,000,000 prize, but anyway. You could still give it a shot, just for the fame and glory ...

Why am I telling you this? Because while 64 = 61 + 3 = 59 + 5 = 53 + 11 = 47 + 17 = 41 + 23 can be written as the sum of two primes in 5 different ways and 66 can be written as the sum of two primes a whopping 6 ways (you're on your own there, work it out; you can check your findings at the back ...) 68 can only be written as 61 + 7 and 37 + 31. Currently 68 is the largest known number that can be written as the sum of two primes in only two different ways.

Keeping up with the Joneses

Welcome to 69 Puzzle Street. The house belongs to the Jones family, and it's a big day.

James Jefferson Jones, only son of Jeremiah and Jacinta Jones, is celebrating his birthday with his 5 closest friends, Jamal, Jelani, Jitka, Jozsa and Juanita who are all different ages from 15 through to 19, one of which is the age James is turning today. It should be obvious that the letter J plays a big role in James' life so he decides to throw a J-themed party.

As he looks around the room, he notices that each of his friends is eating a different junk food, while doing a different activity to burn off the joules.

1. Jozsa, the Jazz Dancer, and the Jaffa-eater are all different people
2. Jitka is either eating Jawbreakers or doing Jiu Jitsu
3. The person who is Jogging is doing so while chewing on Jubes
4. Juanita is 17 and eating Jelly
5. The person doing Jefferson Squats is 19
6. Jelani is Jumping Rope
7. Jitka who is 16 is noshing down on a Jam Doughnut
8. Jamal is 18

James is the same age as the person chewing jubes. How old is James, and who is his friend of the same age?

Economy class passengers have a 69% survival rate in plane crashes. First class? Only 49%

Popular Mechanics measured the data from 20 aviation accidents in the United States from 1971 until 2017 and found that despite paying much more for their tickets, getting the fancy fully-horizontal beds, Champagne upon arrival and those really cool comfy slippers ... people in the front section of a plane were almost one-and-a-half times as likely to die in a crash.

At the same time, only around 1 in every 3,000,000 flights ends in a fatality, so if you've got the option of a complimentary upgrade to the land of silver cutlery and an unending supply of Toblerone chocolate ... I'd take it.

The dust in your house is *not* 70% dead human skin!

It's one of those great urban myths. Yes, you do shed skin — about 1 million of your more than 1 trillion (1,000,000,000,000) skin cells fall off per day, a total of over 4 kilograms in a year — but most of them are washed or scraped off by bathing, shaving, and the like. If you think a trillion skin cells is a lot you have something like 30–40 trillion cells all up in your body. It's actually hard to calculate and estimates vary.

Anyway the dust in your living room contains far more animal fur and feathers, sand and dirt, insect poo and minute food scraps than it does your skin.

It's heptomino to be square (trust me, kids, in 1986 this would have been *hilarious*!)

Just like the lovely little pentominoes we met back at 12, we call shapes made by joining 7 squares by their edges 'heptominoes'.

The name comes from the ancient Greek word 'hepta' meaning — you guessed it — seven. In Latin they went for 'septem', which is why we celebrate the month of September, not Heptember which, and it might be just me, sounds far cooler.

But, Adam, September is the 9th month not the 7th ... what gives? Well, my curious friend, I'd recommend you race out and buy a copy of my *Big Book of Numbers* which explains all that stuff in great depth.

Okay ... or you could just Google it.

Anyway, the grid over the page contains 70 squares. If you look long enough, you should be able to place the 10 heptominoes I've given you into the grid so that each one of them contains the 7 letters SEVENTY in some order. Some of the heptominoes have to be rotated, but none are reflected from the way I have given them to you.

I've given you one to start, best of luck finding the other 9.

Y	T	N	E	V	E	S	E	S	Y
E	T	**S**	**V**	**T**	**N**	E	V	E	T
N	E	S	**E**	**E**	**Y**	V	E	V	N
Y	V	N	T	Y	V	E	S	N	E
V	E	S	T	Y	Y	N	E	T	V
E	E	S	N	E	N	T	S	Y	E
E	V	E	N	S	T	Y	E	N	S

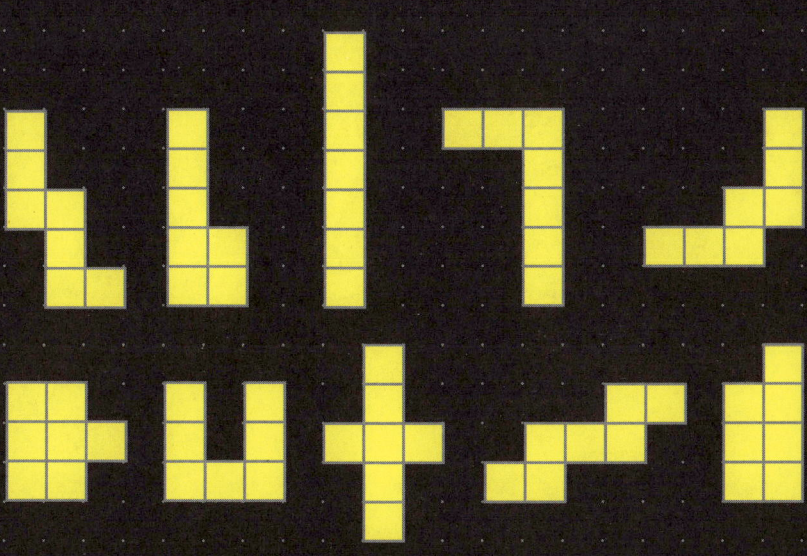

As early as '71, a Spanish doctor suspected a link between smoking and cancer

No, not 1971 … 1571. But although Spanish doctor Nicolas Monardes had correctly perceived a link between the two, he was –cough (pun intended) – slightly out in his findings. He concluded that smoking tobacco could cure more than 20 illnesses including … yep … cancer.

Number theory

The number 71 might seem like a stock standard old prime number to you. Tucked away between the primes 67 and 73, not really hassling anyone.

But 71 is a solution to two fascinating little problems in a branch of mathematics known as 'number theory'. In many ways, number theory looks at the sort of mathematics that non-mathematicians would most think of when they tried to imagine what mathematicians do for a living. For example, number theory problems might concern factorising numbers or determining if they are prime.

Seventy-one is a Pillai prime, named after the Indian mathematician Subbayya Sivasankaranarayana Pillai because $9! + 1$ is divisible by 71 but 71 is not one more than a multiple of 9.

Sounds ugly, but read it a few times and prove it (by hand please — no calculators).

Seventy-one is also part of the last known pair of Brown numbers. Brown numbers are solutions to what is called Brocard's Problem (named after the incredibly coolly named Pierre René Jean Baptiste Henri Brocard); Brocard's problem involves finding numbers n and m that satisfy $n! + 1 = m^2$.

Read it a few times and then find the value of n such that $n! + 1 = 71^2$. Then really impress yourself by finding the other pairs of Brown numbers that involve m less than 71.

Go right ... now

We mentioned the dreaded football penalty shoot-out back at 48. It really is one of the most horrible things to endure in sport. Trust me, I've coached and played in several teams who've been involved in them.

People often say goalkeepers have nothing to lose in a shoot-out. One of them will emerge a hero and the other, well, what could they have done? But keepers obviously feel just as much pressure trying to make the big save as the shooter does trying to find the net. And perhaps that pressure influences the goalkeeper's reactions?

The keeper may decide in advance to wait for the shot to leave the player's boot, hoping that if it is not really well hit they can make a save. Alternatively, they might choose which way they will dive beforehand, giving themselves time to dive all the way to the post ... if they've chosen the right way!

A 2011 study by psychologists from the University of Amsterdam published in the journal *Psychological Science* found goalkeepers dived to the right 71% of the time when their team was losing in the shoot-out, but only 48% when they were leading and 49% of the time when the shoot-out was currently tied.

This subconscious favouring of the right doesn't have anything to do with being left-handed or right-handed which might be the first theory that popped into your mind. But we do know humans, dogs and some other animals, tend to move to the right when they come close to something they really want. For

example, when doing embarrassing displays of kissing in public, lovers tend to lean their heads to the right; similarly dogs have long been known to wag their tails to the right when their masters approach.

So the next time you're about to take that spot kick that could win the European Champions League, block out the roar of the 95,000 fans, the fatigue and the fact that this IS the most important moment of your life, and slip it to the goalie's left. It just might work out for you!

A quick question for you: if team A beats team B 5-4 in a penalty shoot-out and we know that team A took the first shot and we didn't need to go to the 6th shooters from each team, how many ways might the successful and unsuccessful shots have occurred?

There is only one way. If X means a goal and O means a miss, team A has clearly shot X X X X X. At this stage you might be thinking, 'Team B missed one shot, it could have been any of the 5 shots, eg O X X X X, or X X O X X ...' but if you think about the rules of the shootout, O X X X X isn't possible here because when team A scored their 5th goal, they would have led 5-3 and the shoot-out would be over. The only possibility is that team B shot X X X X O. So (X X X X X - X X X X O) is the only possible way the 5-4 shoot-out could be won.

These are harder, but try to calculate how many ways a shoot-out could end 4-3, 4-2 and 3-1. Each time assume team A shot first and that we didn't go beyond the 5 shooters initially nominated by each team. Hint: you have to think about how the actual shoot-out would have played out, beyond just counting all the combinations that end 4-3, and so on.

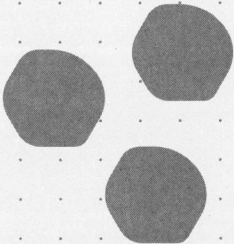

Back at 32 ...

We learned that the largest raindrops can impact the ground at 32 kilometres per hour.

If you think a heavy raindrop is travelling fast, if a large hailstone — say 4 centimetres across — hits you, just imagine what's left of your umbrella or the roof of your brand new car at, you guessed it, 72 kilometres per hour!

I should point out the physics of calculating hailstone velocity is complicated because hailstones aren't perfect spheres, and they melt and bump into other hailstones as they fall. But at the end of the day that doesn't make them hurt any less.

AS inc.

Adam Spencer

Puzzle fanatic

Email book @adamspencer.com.au
Twitter @adambspencer
Instagram adam_spencer1

www.adamspencer.com.au

Take my card ... and hers ... and his ... and ...

A group of complete strangers meet for dinner. If 72 business cards change hands, and everyone swaps one of their business cards with everyone else, how many strangers came to dinner?

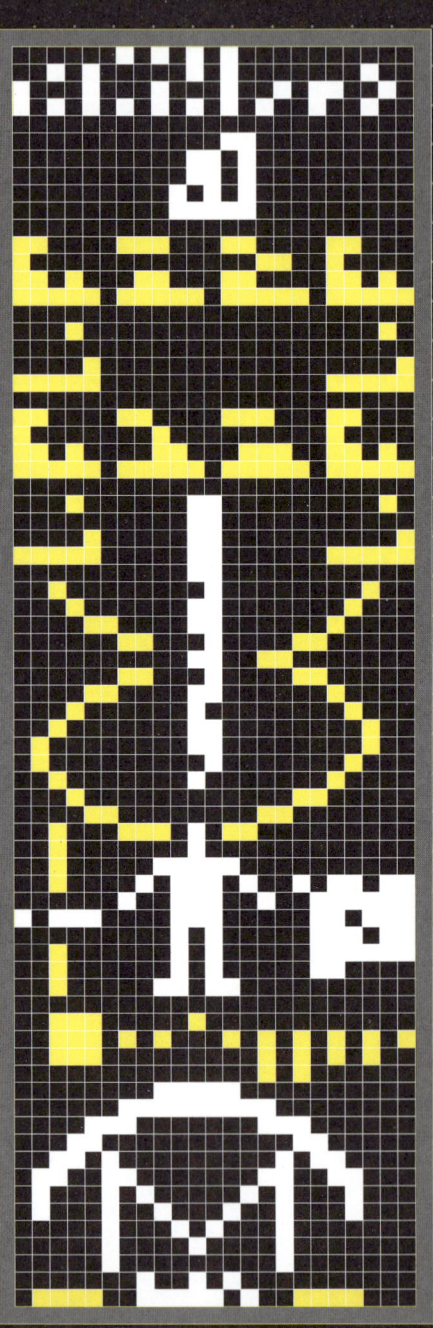

This groovy 73 × 23 grid …

is the famous Arecibo message which was broadcast into space on 16 November 1974 in the hope that an intelligent alien civilisation would hear it and respond.

It contains all the stuff you'd expect, the numbers 1 to 10, information about basic elements, DNA, the solar system and the rough shape of a human.

No one has gotten back to us yet, but please don't think they are being rude. The globular star cluster M13 at which we aimed the Arecibo message is about 25,000 light years away, so even a prompt reply will take a good 50,000 years for us to get. I hope that if we've gone out they have the sense to leave a message.

Party at Gorgeous Gary's Goldshop

Number 73 Puzzle Place is one of most popular houses in the neighbourhood — it belongs to Gorgeous Gary the generous Goldsmith.

Gary is quite a character who loves a good puzzle … and a party. Each year on his birthday he invites all his neighbours on Puzzle Place for a good old get-together, but with a twist.

To get into this absolutely cracker party, you have to answer one of Gary's fiendish logic puzzles. And once inside, there's a chance to walk away with a piece of gold! But again — you guessed it — you have to prove to Gary that you've earned it.

You arrive ready for some serious partying, only to realise the act of entering the party is actually this year's puzzle. There are 3 windows on the front of Gorgeous Gary's Goldshop and alongside them sits the gorgeous one himself.

'Climb through the unlocked window and party away,' he beams, 'but try either of the locked windows and it's back home for another year, my friend.'

The 3 windows are labelled as follows, with a hint on them too:

This window is unlocked	This window is locked	The *left* window is locked
LEFT	MIDDLE	RIGHT

Gary chuckles as you roll your eyes and he says, 'Well, party animal, I can tell you that of these 3 statements at most one is true.'

You climb through the window and into the party! Which window did you choose?

Gorgeous Gary's Gold!

So you made it inside Gorgeous Gary's party? Party on!

The party is now in full swing and, would you believe it, Gorgeous Gary draws your name at random to go in his annual Gorgeous Gary's Great Gold Giveaway.

It's clear that Gary loves his alliteration and his puzzles. Because this time, in the centre of the room, there are 3 suitcases, coloured grey, yellow and white (I know, the exact same colours as we chose for this book — what are the odds).

Gary informs you that one is full of gold and the other two contain ... nothing.

Each case has a note on it, giving a hint as to its contents.

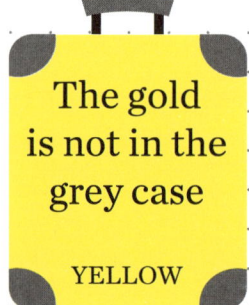

The gold is not in the grey case

YELLOW

The gold is not in this case

GREY

The gold is in this case

WHITE

This time, His Gorgeousness tells you that at least one statement is true and at least one is false and that you have 60 seconds to pick just one case.

You think about it for 58 seconds ... and open the case containing the gold.

Start a stopwatch now and grab that gold.

This puzzle is inspired by a puzzle by one of the greats, Raymond Smullyan, from his epic collection *What is the name of this book?*

If you love puzzles, you must track down a copy of this classic.

74	97	97	74	82	74
82	28	28	82	28	97
82	28	28	82	28	97
74	97	97	74	82	74
97	28			74	97
74	82			82	28

Try to score with 74

The number 74 occurs 8 times in this grid.
Hey, go ahead and count them, if you don't believe me.
If we complete the pattern in the 4 empty squares, do any more 74s occur?
Where?

Forget snakes or spiders ...

According to research from Melbourne University's Dr Ronelle Welton, more Australians have been killed by horses (74 at the time of writing) since 2000 than sharks (26) or bees and wasps (27).

Burns accounted for a whopping 974 deaths in Australia but all were well and truly below the astonishing 4820 deaths attributable to drowning. Always swim between the flags, people.

Oh, and our 8-legged friends, the spiders? Not a single death, I'm pleased to say (although almost 12,000 people were hospitalised).

Some sum!

A popular arithmetic exercise is to try and express the number 100 using all the digits 1 through 9 once each and the basic operations of plus, minus, multiply (times) and divide.

For example, the most commonly quoted example is the beautiful:

$$1 + 2 + 3 + 4 + 5 + 6 + 7 + 8 \times 9 = 100$$

Or, as you hopefully saw when getting the answers to 067, grouping digits together, you can get:

$$123 - 45 - 67 + 89 = 100$$

Or, if you're feeling a bit wild and want to use square root signs as well, how about:

$$(1 + 2) \times (- 34 + 56 + 78) \div \sqrt{9}$$

Anyway, I'm thinking of an expression for 100 that doesn't need square root signs, just the basic operations and concatenation, and it looks like this:

$$100 = 75 + {}^?\!/_?$$

But sticking to the 75 theme, one of those question marks also equals 75. Can you work it out using the remaining digits 1, 2, 3, 4, 6, 8 and 9 only once each?

If you'd like a hint, and why wouldn't you, my solution has 3 digits on the top and uses a multiplication and addition sign and 4 digits on the bottom using a division and a minus.

Can you generate the expression?

Make yourself psychic with the toss of a coin!

Take a room of schoolchildren, get half of them to toss a coin 75 times and write down the result. Get the other half to fake it, and write down the result of 75 coin tosses without actually tossing a coin. Get them to write their names on the top of their sheets and hand them to you.

Watch them marvel as you name them and correctly state whether they tossed for real or faked it!

The secret lies in the fact that almost anyone who is faking will not write down a string of 4 heads in a row, or 5 tails or anything beyond maybe a run of 3 that are the same. A typical fake is HTHTTHHTHTHTHHTHTTTHHT ... To the untrained mind it 'seems fake' to write HTHHTHHHHHHHTHHTTTTH with long strings the same.

But 75 tosses of a coin can be thought of as 71 strings of 5 tosses — toss (12345), (23456), (34567), ... (71 72 73 74 75) and in any string of 5 tosses there is a 1 in 16 chance of seeing HHHHH or TTTTT. So we should expect around 4 runs of 5 Hs or 5 Ts.

Similarly, there is a 1 in 8 chance of HHHH or TTTT in the 72 strings of 4 tosses so it's not unusual to see 9 strings of HHHH or TTTT. (Note here that HHHHHH is 3 strings of 4 overlapping.)

What is very unusual is 75 tosses to contain NO strings of 4 or more Hs or Ts. Name these sheets as the fakers and you'll blow their minds.

(75¢) stamp of approval

The post office on Puzzle Place sells two types of stamps. They are worth 25 cents and 75 cents.

One day Michelle Mailer enters the store, says, 'I've come to buy one stamp' and gives the woman at the counter $1.

The woman immediately hands her over a 75 cent stamp and Michelle says thank you, takes her change and leaves.

How did the woman at the counter know that Michelle Mailer wanted a 75 cent stamp without her mentioning that?

An incredibly rare collection of over 100 rare plant specimens ...

... was held in quarantine in Australia for 76 days ... before a paper-work mix-up saw them destroyed in an incinerator!

The herbarium sheets were on loan from the Museum of Natural History in Paris to the Queensland Herbarium but when they had sat in biodiversity facilities for well in advance of 30 days, and authorities didn't realise that discussions with the Herbarium were still ongoing, they were destroyed which is standard procedure for items deemed to be 'of low value'. It was described as a 'deeply regrettable occurrence'.

987654321 call me ...

For the number 75 we just wrestled with some pretty heavy equations expressing 100 using the digits 1 through 9 each once only.

I also showed an example of sums using 1 to 9 in that order.

Well, it won't surprise you that we can express 100 as a sum of numbers that read 9 to 1 in *descending* order.

For example:

$98 + 7 + 6 - 5 - 4 - 3 + 2 - 1 = 100$ or $9 - 8 + 7 + 65 - 4 + 32 - 1 = 100$ or $-9 + 8 + 7 + 65 - 4 + 32 + 1 = 100$.

There are 18 solutions in total that satisfy this order of 9 through to 1 using only + and – signs, and joining digits into bigger numbers (which we call 'concatenation' as you might recall from number 34).

Why am I telling you this now? Because 7 of them involve the 7 and 6 being joined to form ... 76.

Now, in 6 of these cases it is a +76 so from the numbers 9, 8, 5, 4, 3, 2, 1 in that order you need to create +34.

In one case we use –76, so the remaining digits need to give us 176 (this gives you a hint as to what the 9 and 8 and the 5 and 4 need to do!).

Try to find all 7 such equations and if you're enjoying yourself the other 11 in which the 7 and 6 appear discretely.

Per cent pow!

Last year across Canada, Bacardi had to recall a batch of Bombay Sapphire Gin.

Rather than the usual 40% alcohol, these rogue bottles were packing 77%! Bacardi believes that while swapping bottling tanks during the production process, a faulty valve may have allowed an extra high strength spirit to sneak into some bottles.

Bacardi stressed that it was not potentially fatal but best to be safe and return any possible overstrength bottles.

And on the topic of alcohol consumption. Those brilliant brain beavers the QI Elves point out in the Second Book of General Ignorance *that the member of the animal kingdom that hits the sauce most after humans is ... the Malaysian Pen-tailed Shrew.*

The shrew, known as Ptilocercus lowii, (try saying that after a couple!) imbibes alcohol from the nectar of the Bertram Palm tree and downs an average 9 units of alcohol per night. That's the equivalent of a bottle of wine.

Not bad for an animal the size of a rat. That's what I call getting completely 'Pen-tailed shrew'-faced.

The kormas at no. 77

Achill breeze blows up Puzzle Place reminding our good friends Aaron, Barry, Caroline, Danielle, Eamon, Fantine, Gary, and Hayley that winter is well and truly here.

That is why they have chosen Indira's Indian Institution for their latest ingestive incursion.

As they enter 77 Puzzle Place and make their way to the table, Aaron, being his usual controlling self, sits in seat number 1 and asks Fantine to sit opposite him. He then arranges everyone as per his particular wishes. Once they are seated it turns out that:

- Danielle sits opposite a female
- Eamon sits closer to Fantine than he sits to Caroline
- Barry and Hayley sit the same distance from Gary, with Barry closer to Gary's right
- There are an odd number of chairs between Caroline and Hayley
- Barry, Danielle, and Fantine all sit together in some order

So, my dining detective, can you label the seats in the way our 8 friends were arranged? Good luck and see you again in spring!

Since 1985, 78 is the number of laps in the Monaco Grand Prix — except 1989...

... *when it was 77 laps. I have no idea why it was and might have to ask my sister Leisha who is an absolute F1 nut.*

Nevertheless, since 1985, the Monaco Grand Prix has consisted of 78 laps around the famous streets of Monte Carlo and La Condamine.

Nelson Piquet once described navigating the beautiful but demanding harbourside street circuit based in the playground of the rich and famous as 'like trying to cycle around your living room'!

The final count-up

So we've now had a fair crack at taking the numbers 1 to 9 in numerical order, concatenating (joining) some of them and making equations using addition and subtraction to get 100.

For example, the case involving the concatenation 34

$$1 + 2 + 34 - 5 + 67 - 8 + 9 = 100$$

... from back at, you guessed it, 34. This is also one of the 5 such equations that involves 67, which is the most common concatenation in all the equations of this type. It turns out, the 12 possible sums involve 67 five times, 89 four times and our buddy 78 pops up on three occasions.

You don't even need me to ask, do you ... off you go and find the 3 such sums involving 78. If somehow you've landed on this page first up without doing the puzzles for 34 and 67, go back because they give you a good toolbox of strategies to significantly simplify what you need to do here.

About those perfect number ...

A perfect number is equal to the sum of its proper divisors. So the divisors of 28 are 1, 2, 4, 7, 14 and 28 and we see that $1 + 2 + 4 + 7 + 14 = 28$. This makes 28 perfect. The smallest perfect number is 6 (note that $1 + 2 + 3 = 6$).

A number is 'semi-perfect' if it is the sum of some or all of its proper divisors.

So the proper divisors of 20 are 1, 2, 4, 5 and 10. Dropping the 2, we see that $1 + 4 + 5 + 10 = 20$. So 20 is semi-perfect.

Some handy facts about semi-perfect numbers include:

A multiple of a semi-perfect number is semi-perfect, so 20 being semi-perfect tells us that 40, 60, 80, ... are too.

If no divisor of a semi-perfect number is itself semi-perfect we call that number primitive. You can check that 20 is also a primitive semi-perfect.

All perfect numbers are semi-perfect, because they are the sum of ALL of their proper divisors.

So all multiples of perfect numbers are semi-perfect. This tells us that 78 being equal to 6×13 must be semi-perfect. And sure enough, out of its proper divsors 1, 2, 3, 6, 13, 26, and 39, we can see $78 = 13 + 26 + 39$.

And it is! The proper divisors of 78 are 1, 2, 3, 6, 13, 26 and 39 and $13 + 26 + 39 = 78$. Note how this is just the equation $1 + 2 + 3 = 6$ with every term multiplied by 13.

Down low, too slow

A group of adults and children go to lunch.
The adults all shake hands, as adults do, but the boys all give each other high fives while the girls exchange fist bumps (frightfully cool little things, aren't they!). While adult men and women might shake hands, obviously boys and girls don't come into contact with each other!

In total, 78 handshakes, high fives or fist bumps occur. If there were both more boys and more girls than adults at the lunch, how many adults attended?

Seventy-nine is prime

Remember from school that 6 is NOT prime because we can write 6 = 2 × 3, whereas 7 is prime because we can write 7 = 1 × 7 but cannot break it down into any smaller factors. We call numbers like 6 composite numbers.

So here are a few amazing prime time facts about 79!

Being prime, we know 79 must be an odd number because apart from 2, which is prime, all even numbers are divisible by 2 and hence composite.

In fact, 79 is the 22nd prime number, nestled between 73 and 83.

79 is an emirp (prime spelt backwards) because its reverse 97 is also prime. We also call 79 and 97 permutable primes.

79 is a cousin prime with 83 (because they differ by 4) and a sexy prime with 73 (because they differ by 6).

79 is a Pillai prime because 79 is a factor of 23! + 1 while 79 is not one more than a multiple of 23.

79 is also a fortunate prime, a happy prime, a Higgs prime, a Kynea prime and a Gaussian prime. While you're there, you can add lucky, regular and right-truncatable prime.

And to cap it all off, 79 is the smallest number that cannot be represented as a sum of fewer than 19 fourth powers. . This is pretty special, because in 1909 mathematical gun David Hilbert proved that every natural number is the sum of at most 19 fourth powers. The breakdown for 79 is that it's made up of 4 lots of 2^4 plus 15 lots of 1^4.

79, you're looking fine!

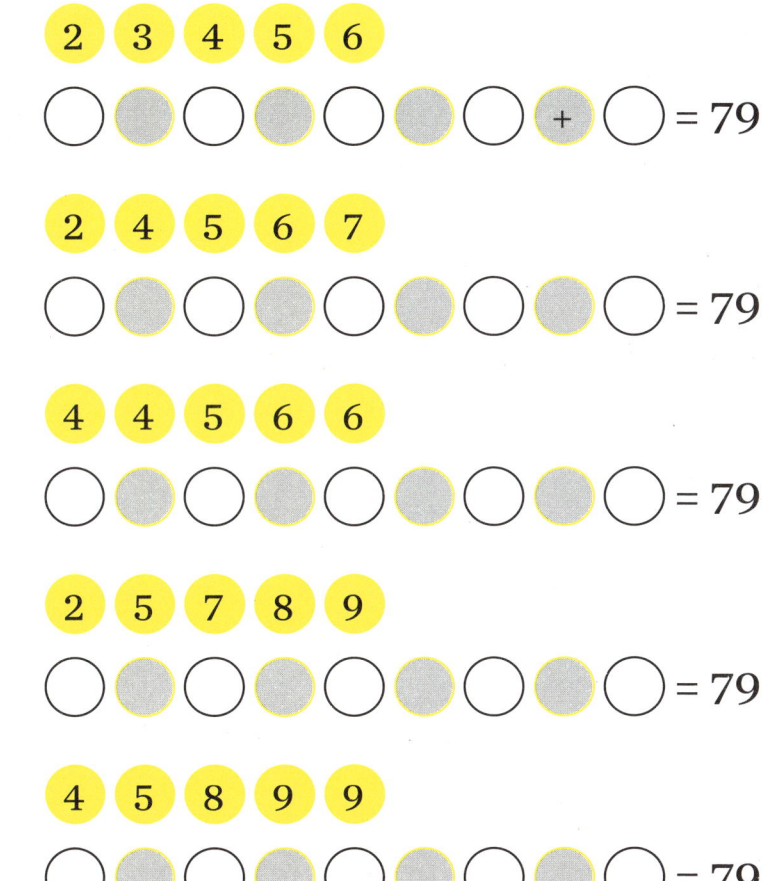

2 3 4 5 6

◯ ◯ ◯ ◯ ◯ ◯ ◯ + ◯ = 79

2 4 5 6 7

◯ ◯ ◯ ◯ ◯ ◯ ◯ ◯ = 79

4 4 5 6 6

◯ ◯ ◯ ◯ ◯ ◯ ◯ ◯ = 79

2 5 7 8 9

◯ ◯ ◯ ◯ ◯ ◯ ◯ ◯ = 79

4 5 8 9 9

◯ ◯ ◯ ◯ ◯ ◯ ◯ ◯ = 79

Sum, er, of 79

Y ou know the game.
But suffice to say, these little equations are even tougher.

Once again, you're trying to find the arrangement that gives you the highest 'score', which you get by reading off the yellow numbers in order.

I'll start you off with a single operator in the first equation, but after that, it's just you, your 2B pencil and your little grey cells!

Hidden between the lines

Since 2010, adventure lovers and puzzle sleuths from around the world have searched America's Rocky Mountains hoping to find the Fenn Treasure — a collection of gold and gems worth over $2 million buried by art dealer Forrest Fenn.

Fenn himself was good enough to write a 24-line poem that is said to contain 9 clues as to the whereabouts of the bounty.

As I have gone alone in there
And with my treasures bold,
I can keep my secret where,
And hint of riches new and old.

Begin it where warm waters halt (1)
And take it in the canyon down, (2)
Not far, but too far to walk. (3)
Put in below the home of Brown. (4)

From there it's no place for the meek,
The end is ever drawing nigh;

There'll be no paddle up your creek, (5)
Just heavy loads and water high. (6)

If you've been wise and found the blaze, (7)
Look quickly down, your quest to cease,
But tarry scant with marvel gaze, (8)
Just take the chest and go in peace.

So why is it that I must go
And leave my trove for all to seek?
The answers I already know,
I've done it tired, and now I'm weak.

So hear me all and listen good,
Your effort will be worth the cold.
If you are brave and in the wood (9)
I give you title to the gold

Happy hunting, but be careful. At least one person has died in the rugged wilderness. Here, Forrest Fenn offers a final clue: the prize isn't 'any place where an 80-year-old man couldn't put it'.

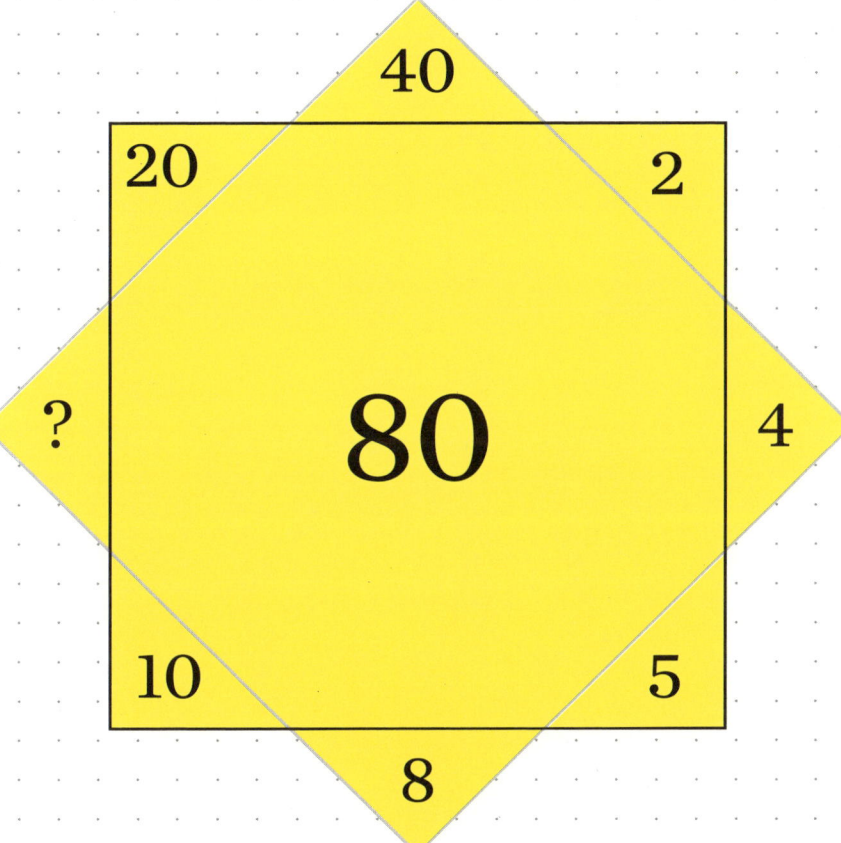

Earn yourself a gold star

Well, the question's simple ... ... but, how about the *answer*? What number replaces the question mark in the diagram above?

You acting the goat, sunshine?

On 16 June 2006, at a military parade in Cyprus to celebrate Queen Elizabeth II's 80th birthday, Lance Corporal William Windsor of the Royal Welch Fusiliers refused to keep in line, ignored orders and even attempted to headbutt a military drummer. Following a disciplinary hearing Windsor was demoted to fusilier.

The only thing that could be said in his defence was that William Windsor ... was a Kashmir goat! Since 1844 the British monarchy has supplied Kashmir goats to march at the head of the Fusiliers on ceremonial occasions.

He was reinstated to Lance Corporal in September of 2006 and on 20 May 2009, following 8 years of distinguished service, William Windsor, army number 25232301 retired and was replaced by William Windsor the Second.

24 & 26

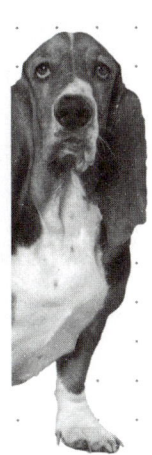

Gematria redux

Remember back at 14 we married the classical musician Bach with the code a = 1, b = 2, c = 3 ... z = 26?

Adding the sums of the letters in words you can see that CAT = 3 + 1 + 20 = 24 and DOG = 4 + 15 + 7 = 26.

You can even find that ZERO = 64 and ONE HUNDRED = 34 + 74 = 108.

Playing this game, what is special about the number 80?

You might want to try several numbers near 80 to see if the answer spells

Shogi, mate

There are 81 squares on a board for the Japanese game of Shogi, sometimes called Japanese chess or the General's Game. Similar to chess, the aim is to trap your opponent's King which is surrounded by Pawns, Knights, Bishops and Rooks but also Gold and Silver Generals and Lances.

Eighty-one fun (though a nasty one!)

Now we all know that 100 = 81 + 19.
Nothing impressive there, Adam. But this is really cool.
You can write:

$$19 = \frac{\bigcirc\bigcirc\bigcirc\bigcirc}{\bigcirc\bigcirc\bigcirc}$$

... using the remaining digits 2, 3, 4, 5, 6, 7, 9 that aren't used in the number 81. So we need a fraction using these digits that is equal to 19.

There are actually two fractions that satisfy this condition. You can smash it out by trial and error, but I'll help you with some hints to narrow down the options. Both fractions obviously have 4 digits on the top and 3 on the bottom. The 9 appears somewhere in the bottom of each fraction, the 4 and the 5 are always on the top. Further, the 6 and the 7 do not appear together, one is on the top and one on the bottom each time. The same for the 2 and the 3 which never appear on the same level of the fraction. But the 2 and the 7 are on the same level each time as is the 3 and the 6.

Huh? Try to write out a long multiplication table like this:

... and have a bash. Good luck.

					=	22
	■		■		■	■
					=	49
	■		■		■	■
					=	28
=	■	=	■	=	■	■
56	■	7	■	72	■	■

Grid 1 (left):

82		28		2	=	68
	■		■		■	
2		8		28	=	38
	■		■		■	
8		2		2	=	32
=	■	=	■	=		
33		24		58		

Grid 2 (right):

	—		—		=	52
—	■	×	■	÷	■	
	×		÷		=	32
—	■	—	■	+	■	
	×		+		=	98
=	■	=	■	=		
66		14		96		

Thanks for nothin' ...

The grids here, when completed, contain combinations of the number 82 and other numbers made out of its digits, namely 2, 8 and 28.

These numbers are separated by + − × and ÷ signs, and the equations formed by each row and column give the answer at the end of the row or column.

As always, don't forget that order of operations applies. So for example, $28 - 8 \times 2 = 28 - 16 = 12$.

In the first grid above, I give you the numbers and need the operator signs, + − × and ÷ sign.

In the second, harder puzzle on the right, I give the signs and need the numbers.

And on the opposite page ... well, it's eye-bleed time as I give you nothing at all but the answers to each equation.

Have fun!

If you love American sports you're probably familiar with the number 82

Both the NBA basketball and NHL hockey leagues see teams play an 82-game regular season before moving into the 'play-offs'.

If you think 82 games is a long slog, spare a thought for major league baseball players. The 30 MLB teams play 162 games in their regular season but still 82 is significant. If you have 82 or more wins, you've had at most 80 losses and therefore a winning season.

$$18 = \frac{\text{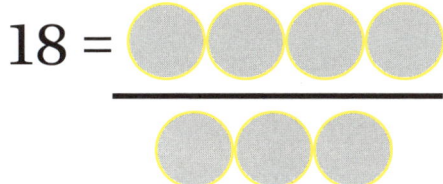}}{\text{ }}$$

Plenty to do ...

Similar to the examples that we saw for 81, we can write 100 as the sum of two numbers: 82 plus a fraction where the fraction involves the other digits 1, 3, 4, 5, 6, 7 and 9.

So, again you want the fraction on top above, which equals 18 and will fit into the multiplication algorithm.

For 81 you just found two solutions, but for 82 there is only one. Once again, it has to fit the multiplication algorithm using the digits 1, 3, 4, 5, 6, 7 and 9.

Need a hint? The top line of the fraction (the numerator) uses 4 consecutive digits, but not necessarily in numerical order, like, say, the number 1423 or 8576.

This is a good hard question to crunch out and requires that you really focus. Good luck!

Miwin's amazing non-transitive prime-numbered dodecahedra

Way back at 24 and 25 we met examples of 'non-transitive dice'. These are dice with different numbering to the traditional 1 to 6 and where when rolled against each other, for example with dice A, B and C, over time A will beat B, B will beat C, but C will beat A.

Well, it wouldn't be one of my books without a mind-blowingly difficult example of an earlier concept brought back near the end of the book. Meet Miwin's nontransitive prime-numbered dodecahedra! These dice, invented by physicist Michael Winkelmann in the 1970s, each have 12 pentagonal faces and the numbers on every face are primes, with faces ranging from values of 7 to 83.

Winkelmann discovered two sets of 3 dice each. In each set, any pair you roll against each other will see one die win over time in a ratio of 35:34. But within each set A beats B, B beats C and C beats A making the dice non-transitive.

Above, you can see a typical Miwin non-transitive, prime-faced dodecahedron net ... just chillin'.

Prime real estate

Today, 2 September 2018 at number 83 Puzzle Place, is a very special day. One of Puzzle Place's nerdiest residents, Epsilon Erasmus, is celebrating Father's Day with his much younger wife Delta and their three children Alpha, Beta and Gamma. There is a beautiful lunch, complete with cake and presents chosen with Epsilon's love of mathematics in mind.

Epsilon stands to his feet and after saying a few words about how happy he is and how much he loves his family, concludes:

> *'It wouldn't be a celebration in the Erasmus house without me making a mathematical observation. I love prime numbers. It's part of why we live at number 83 Puzzle Place. This year is 2018, and of course 2018 is not a prime number. The next prime year will be 2027, and the 10-year stretch of years without a prime, from 2017 until 2027, is indeed a long one.*
>
> *To find a further stretch you have to go back to the years 1951 until 1973, a 21-year stretch of non-primes during which I was born, but your mother wasn't … no, she just missed out.*
>
> *Speaking of your mother and I, what is particularly lovely about this year is that if you add up your ages, Alpha, Beta and Gamma, you get your mother Delta's age, and like the year 2018, none of those ages are prime. But on this day in the next prime year, 2027, you can add your three ages and get* my *age — and all of those ages will be prime numbers. In fact, your ages, my sweet little things, will be consecutive primes.'*

Epsilon Erasmus sits down with a geeky grin on his face. What is the age difference between Epsilon and his wife Delta?

A second-hand pair of Converse sneakers recently sold for $US190,373

They were the shoes none other than Michael Jordan wore in the gold medal game against Spain in the 1984 Olympics. They were also the last sneakers Jordan wore as an amateur, as well as the last time he wore his trusty Cons in an official game.

Months later, he would be drafted to the Chicago Bulls ... and ink a fairly lucrative deal with a company that goes by the name of Nike. Heard of it?

Actually, on the subject of 84, that year's NBA draft is considered to be one of the greatest of all time, producing no less than 5 Hall of Famers and 7 all-stars.

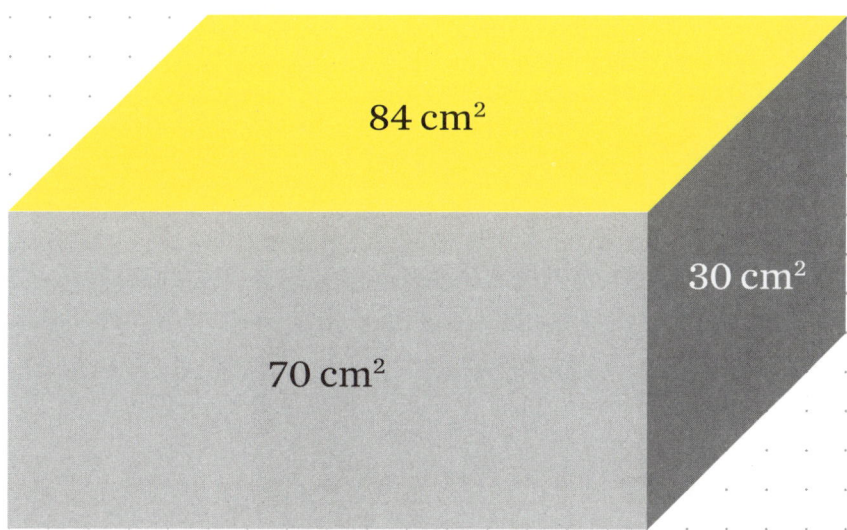

84 cm²

30 cm²

70 cm²

The shoebox problem

The shoebox above has the area of each face written on it.
All the sides are whole numbers of centimetres in length.
What is the volume of the box?

Repent-agon

You may remember the two puzzles using this diagram that we did way back at 14 and 19.

The challenge I set was to arrange the numbers 1 to 10 in these circles so that each triple of circles along a side of the pentagon added up to the same amount. In the cases of each side equalling 14 or 19, there was only one solution and once you've shared the observations I gave you, it wasn't that hard.

Well, this puzzle *is* 'that hard'. This time I want the sides to all equal 17. I'll walk you a fair way through the logic of getting the two possible solutions and then ... yes you guessed it ... it's over to you.

By the same logic we employed earlier, the 5 sides adding to 17 come to a grand total of this puzzle's number — 85. The numbers 1 to 10 add up to 55, so the corners, which we count twice, must give us the extra 30. So for the corners, we need 5 numbers from 1 to 10 that add up to 30.

But notice something else here. If we have 3 of the corners adding up to 17 ... we're stuffed. Just say our corners contained 4, 5 and 8. Try to position these 3 numbers in the 5 corners and you'll see that you always have to have, say, the 5 next to the 8, which would require the 4 in between. Yet we've already allocated 4 to another corner! Or, the 4 is in the corner next to the 5 and we need the 8 in between ... but it's already taken.

So within our 5 numbers that add to 30, we can't have a 'sub-triple' that adds to 17. This is the same as saying the other pair of numbers in the corners can't add to 13. So we can't have 3 and 10 or 4 and 9 or 5 and 8 or both 6 and 7 featuring in our 5 corners.

This reduces the possible groups of 5 numbers greatly. Let's work through them logically.

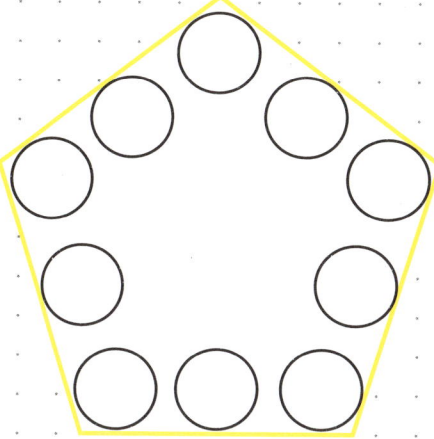

The group of 5 numbers for the corners has to have a largest number. If that is 8, the only possibility is 8, 7, 6, 5 and 4. We can exclude this because it contains 8 and 5 meaning that 7, 6 and 4 add to 17 and can't all be in corners. So the group of 5 numbers has to contain a 9 or a 10.

Let's look at all the possibilities where the largest number is 9. What is the biggest possible sum we can make if we don't include 10? Well, we can start by including 9, 8, and 7... but then we can't use 6, 5, or 4 because there will be a pair of numbers in our set adding to 13. So the next two biggest available numbers to include are 3 and 2. But 9+8+7+3+2 = 29, so even the biggest possible sum we can make is not big enough to reach 30. So our set must include the number 10.

If 10 is included in our set, 3 can't be (since 10+3=13), so what's left to do is find four numbers from the list 1, 2, 4, 5, 6, 7, 8, 9 so that they add to 30-10=20... and don't forget that no two numbers in the group of four are allowed to add to 13. You should be able to find 3 such groups of four... but you still need to check if you can complete the pentagon for each group.

Then if you're feeling really uppity, start from scratch and find the solutions to the pentagon puzzle where the sides all add to 16, AND if you're feeling super zenzizenzizenzic uppity (see 8) prove there are no solutions for the cases of the sides adding to 15 or 18.

x86

Just when you thought understanding computer operating systems, 32-bit vs 64-bit and the like couldn't get any more difficult, get this.

The 'instruction set' is basically the language that a computer will understand. The instruction set for 64-bit computer systems is called x64. But the instruction set for 32-bit programming is called, not x32, but x86.

This is not done just to confuse us, but because the Intel chips that ran 32-bit programs were numbered 8086, 80486 and other numbers ending in, you guessed it, 86.

YOU
ARE
HERE

Newbie of Puzzle Place

There are 86 houses in this part of of Puzzle Place. Felipe First lives in number 1, and Lada Last lives in number 86.

One day Felipe is leaving his home when he notices a new face walking down the street. He strikes up a conversation, and it turns out this is Nadja Newbie who has just moved onto Puzzle Place. Felipe loves his logic problems so, at the risk of sounding like a complete weirdo, says, 'Don't tell me where you live, let me ask you some questions about the number of your house, and I'll see how quickly I can work it out.'

Truth be told, Nadja finds this a bit weird but, hey, she's new to town so she plays along.

Felipe asks Nadja the following questions and she answers with a simple yes or no.

1. Is your house an odd number?

2. Is it a perfect square?

3. Is your house number closer to mine than it is to Lada's?

After just these 3 questions, Felipe says, 'Aha! I know exactly where you live!'
Which house has Nadja moved into?

On the topic of shoelaces ...

Here's a way to tie your shoelaces so that they never come undone.

I discovered this gem when visiting fieggen.com. The site contains 18 different ways to tie your laces, including the Turquoise Turtle, The Surgeon's Knot and the knot that changed my life, Ian's Secure Shoelace Knot. This is a must for busy people on the go, and especially the parents of kids who play football and seem to spend half the game trudging to the sidelines with that empty 'mummy please help me' look on their faces.

It's beautiful. You need laces long enough to be able to create two loops and all you do is rather than pass just one loop through the middle, you pass them both through. One from the front and one from the back. Try it right now. The little 'double wrap' you see in the centre of the knot can practically never undo itself, yet it comes undone in your hands with a firm tug. Woohoo!

Ian Fieggen loves his shoelaces — and even has testimonials to his knot on his webpage. Add me, Ian, this knot rocks!

Laces sold separately?

So we've figured out the dimensions of our shoebox, back at 84. Well, let me tell you about the shoes that came inside.

The cost of the shoes and a pair of laces was $86.

The shoes cost $85 more than the laces.

Jigsores

Remember these babies from numbers 29 and 66?

Well, you should know the drill. Fit the yellow pieces on the right of each puzzle into the grid on the left. Note that the yellow pieces have been scaled down to fit on the page — you're still after one number per grid square on the puzzle.

Remember that your goal is to match each number to its twin. No surprise, these are much, much tougher than the last ones. There are no reflected pieces, but there are rotations. And, yep, the numbers add up to ... the magic number 87.

Let the games begin!

1					

Pieces:

1	1
3	4

1	2
2	3

1	2
3	4

1	2
4	3

1	3
2	4

1	3
4	2

1	4
4	2

1	4
2	3

2	2
2	3

2

| 1 1 | 1 2 | 1 2 |
| 2 3 | 3 4 | 4 3 |

| 1 3 | 1 3 | 1 3 |
| 2 4 | 3 2 | 4 2 |

| 1 4 | 1 4 | 1 4 |
| 2 3 | 3 2 | 4 2 |

3

| 1 1 | 1 1 | 1 2 |
| 3 4 | 4 3 | 2 3 |

| 1 2 | 1 3 | 1 3 |
| 4 3 | 2 4 | 4 2 |

| 1 4 | 1 4 | 2 3 |
| 2 3 | 3 2 | 4 2 |

Eighty-seven is way cool!

It's is the sum of the squares of the first 4 primes ($87 = 2^2 + 3^2 + 5^2 + 7^2$).

It's the sum of the sums of the divisors of the first 10 positive integers ($87 = 1 + (2 + 1) + (3 + 1) + (4 + 2 + 1) + (5 + 1) + (6 + 3 + 2 + 1) + (7 + 1) + (8 + 4 + 2 + 1) + (9 + 3 + 1) + (10 + 5 + 2 + 1)$).

It can be written beautifully using factorials as $87 = 5! - 4! - 3! - 2! - 1!$

And completely unconnected to all of that is ... Swiss Army Knives!

You've probably heard of the handy contraption. The standard model comes with a knife blade, nail file, bottle opener, can opener, a corkscrew and an awl (a piercing tool). But the biggest of all Swiss Army Knives contains 87 tools which can do an incredible 141 functions.

Compared to the normal handy, carry-it-in-your-pocket Swiss Army Knife, the Wenger giant clocks in at 9 inches (23 centimetres) wide, weighs 32 ounces (just under a kilogram!) and retails for a lazy $2100!

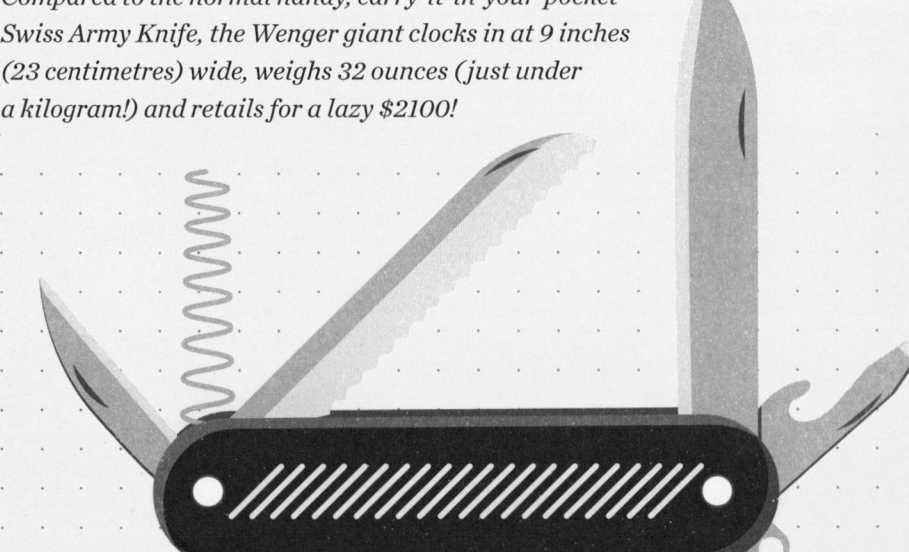

Oh, and 87 = 44 + 43 = $44^2 - 43^2$

Wow, that looks pretty impressive, doesn't it? What a coincidence with the 44 and the 43 coming together like that.

But check this out:

5 + 4 = 9 and $5^2 - 4^2 = 25 - 16 = 9$
12 + 11 = 23 and $12^2 - 11^2 = 144 - 121 = 23$
100 + 99 = 199 and $100^2 - 99^2 = 10,000 - 9801 = 199$

So maybe it's not all that freaky.

In fact, if you remember from high school how to open up brackets with algebra (oldies, if you don't remember it's cool, I'll do the heavy lifting here — just nod along sagely). The good old phrase 'difference of two squares' might sound familiar. That's the rule that tells us the difference of two squares can be expressed as the product of two terms like so: $a^2 - b^2 = (a + b) \times (a - b)$.

In the cases we're looking at, the two numbers a and b have a difference of 1, so the term (a – b) = 1. That means the expression just becomes $a^2 - b^2 = a + b$, which is exactly the pattern we found!

So for any two numbers that differ by one we will see this pattern.

If you wanted to get all mathematical in your lingo you could drop this: 'The difference of the squares of consecutive numbers is always equal to the sum of those numbers.'

$$88^2 = 7744$$

... which looks kind of cool because the 7s and the 4s repeat so 7744 is a square in which there are no 'isolated digits'.

There are actually infinitely many numbers for which this occurs.

The next one after 7744 is 5,500,002,244 but before you go trying to work out the square root of this 10-digit brute, a word of warning. I tried it myself once and found out pretty quickly that 5,500,002,244 = $2^2 \times 11^2 \times$ 11,363,641. But you've got to be pretty patient to find the square root of 11,363,641.

It turns out that 11,363,641 = 3371^2 so the square root of our original number 5,500,002,244 is 2 × 11 × 3371 = 74,162. Yep, that's an hour of my life I won't be getting back!

If you've got many more hours up your sleeve and nothing better to do, why not confirm these calculations, too?

The next few numbers are 105462^2 = 11122233444, 2973962^2 = 8844449977444, and 31500088^2 = 992255554007744.

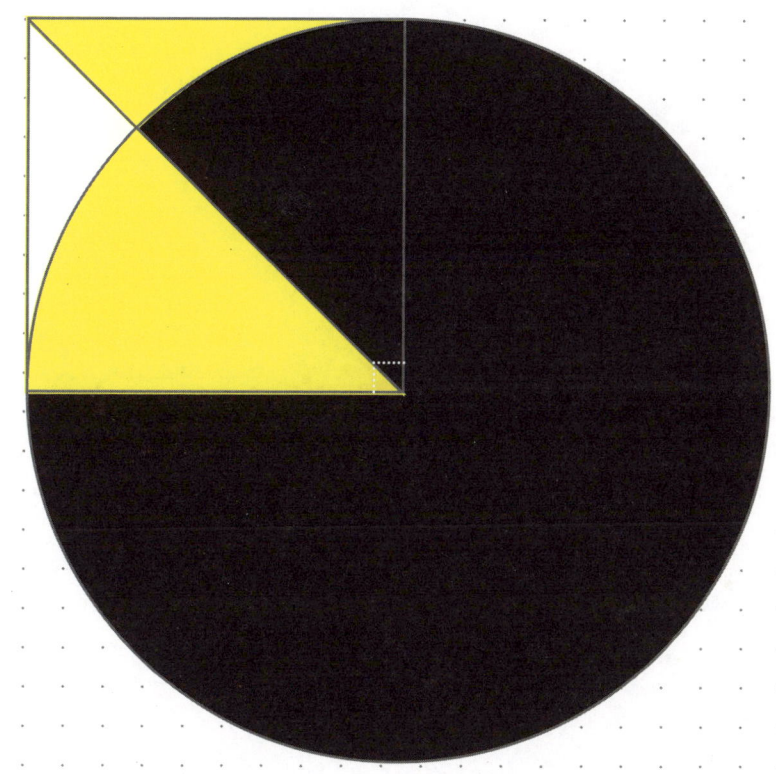

Stands to region

Tell me, what's the area of the two yellow regions combined? All you need to know here is that the radius of the circle is 88 centimetres. Grab your pencil and get calculating!

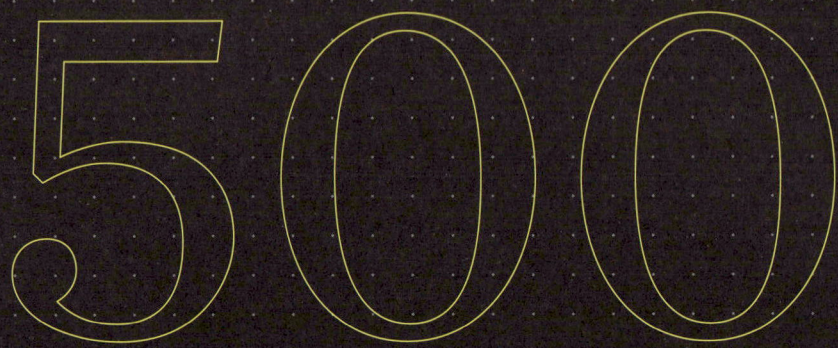

88% of the original Fortune 500 companies ... are no longer!

First published in 1955, the 'Fortune 500' is a list of the 500 largest corporations in America judged by the size of their revenue.

Sixty years after the first list, only 61 companies remained in the top 500. Mergers, bankruptcies and the rapid evolution of modern business in the digital age has seen 88% of these wealthiest firms be replaced ... or vanish entirely.

If you're curious, some of the stayers include Boeing, Campbell Soup, General Motors, Kellogg's, Procter & Gamble, John Deere, IBM and Whirlpool.

... Still stands to region

Okay, this shaded region problem's a shade tougher (ha!).
Each circle above has a radius of 88 centimetres. (Hey, I've scaled them, okay?) So what is the area of the beautiful yellow region?

The geometry needed to get an exact answer may be a bit beyond you, but even if you don't know how to calculate the answer, can you see the beautiful middle step that involves drawing some more lines into this diagram to show you exactly what you should be working out?

Why not have a crack and see how you go.

Hellin's Law

When it comes to the science of childbirth, Hellin's Law is the theory that roughly one in 89 natural pregnancies will result in twins.

Hellin's Law also suggests that triplets pop along on about one in $89^2 = 7921$ occasions. According to Hellin's Law, quadruplets occur ... you guessed it, one in every $89^3 = 704,969$ natural pregnancies. Hellin's Law isn't a hard and fast rule but more a handy 'rule of thumb'.

Another interesting thing about twins, according to the Weinberg differentiation rule, about ⅓ of twins are identical (like the awesome Leo and Jim Carlton who helped proofread this book), ⅓ are fraternal same-sex (which means they don't look identical), and ⅓ are opposite sex.

The curious case of the number 1089

This beautiful little fact was pointed out to me by an awesome young maths geek who emailed me at book@adamspencer.com.au and said I really must include this in my next publication. I'll be honest, awesome young maths geek, my inbox is a jungle and I can't for the life of me find the original email ... but please hit me up again so I can thank you properly!

While we're on the topic, if there's anything you'd like to suggest for a future book, or any questions you've got about stuff in this one, just drop me an email. But I digress.

Check this out. If I take the number 652, reverse it and get 256, subtract the smaller from the larger I get 652 – 256 = 396. Now add that number to its reverse and you get 396 + 693 = 1089.

So what, you say. Let's try it again with another random 3-digit number, say 704. 704 – 407 = 297 and 297 + 792 = 1089.

Coincidence? Nope.

In fact, take any 3-digit number with a different first and last digit, follow this algorithm and you will end up with 1089. If you do enough examples you should start to see why. Note, if the subtraction gives you 99, write that as 099 and then 099 + 990 = 1089.

You might have noticed that if the first and last digit *are* the same, the process just gives you 0. So there are only two possible numbers you can get if you apply this algorithm to any 3-digit number.

Okay, now try the same algorithm out on some 4-digit and 5-digit numbers. How many different possible answers do you think there are in each case?

The final, final count up

Hey Adam, I thought you said 78 was the final count up!

Yes, but we've had so much fun so far investigating the 12 equations that use the digits 1 through 9 in numerical order to give us a sum of 100. Of course you are allowed to join numbers together, as in one of the answers for 78, namely

$$1 + 23 - 4 + 5 + 6 + 78 - 9 = 100.$$

In fact, it should be obvious to you by this stage in the book that we have to do some concatenating because just adding 1 through 9 alone gets you to 45 which leaves us well short of our target number 100.

As I mentioned back at 78, the concatenation 89 occurs in 4 equations of this form. Try to find them. In fact, while you're there, try to discover the entire list of 12 from the beginning again.

And as a special bonus, if I further allow the use of a decimal point between two digits — so for example 12 can become 1.2 — there is exactly one more example which works. As a hint, it uses the concatenation 89 and two decimal points. This sounds horridly complicated but actually if you think about your options with the decimals such that they add to a whole number, you don't have that many options at all.

Off you go, charger!

Fibonacci fun

In 1972 mathematician Ira Gessel, writing in the journal *Fibonacci Quarterly* (where else would you write up an awesome result about Fibonacci numbers?) revealed this beautifully simple test for whether a given number n is a Fibonacci number.

A positive integer n is a Fibonacci number, if and only if $5n^2 + 4$ or $5n^2 - 4$ is a perfect square.

Given that Fibonacci numbers have been around since, well since the time of Fibonacci, who wrote about them in his classic page turner *Liber Abaci* in 1202, it's amazing that it took 770 years for such a simple result to be discovered.

Many of you may not be familiar with the phrase 'if and only if'. It is used to make something both a necessary and sufficient condition for something else. Okay, Adam, that's not helping. Instead let's look at an example. If I said to you, 'I will go to your party if and only if Yana is there,' what does that tell you?

Well, for starters I clearly like Yana but let's leave that for the moment. It is telling you two things. That if Yana is at the party I will go *and* that if Yana is *not* at the party I *will not* go.

If I simply said, 'I will go to your party if Yana is there' I might still be willing to turn up if she doesn't. And if I said, 'I will only go to your party if Yana is there' I might still have other demands as well before I'll turn up. But by saying 'if and only if' she is there I am making the two statements 'Yana is at the party' and 'I will go to the party' effectively equivalent.

So when it comes to positive integers, n being a Fibonacci number and either $5n^2 + 4$ or $5n^2 - 4$ being a square are equivalent statements.

For the largest Fibonacci number below 100, that is 89, we see that $5 \times 89^2 + 4 = 39,609$ and $5 \times 89^2 - 4 = 39,601$ and it turns out that $39,601 = 199^2$.

Convince yourself that for the Fibonacci numbers, $n = 1,1,2,3,5,8,13,21,34,55,89, \ldots$ either $5n^2 + 4$ or $5n^2 - 4$ is a square.

For example, as we just saw, for $n = 89$, $5 \times 89^2 - 4 = 199^2$. The results for small Fibonacci numbers are easy to see. Push this as far up the list of Fibonacci numbers as you can before your brain explodes.

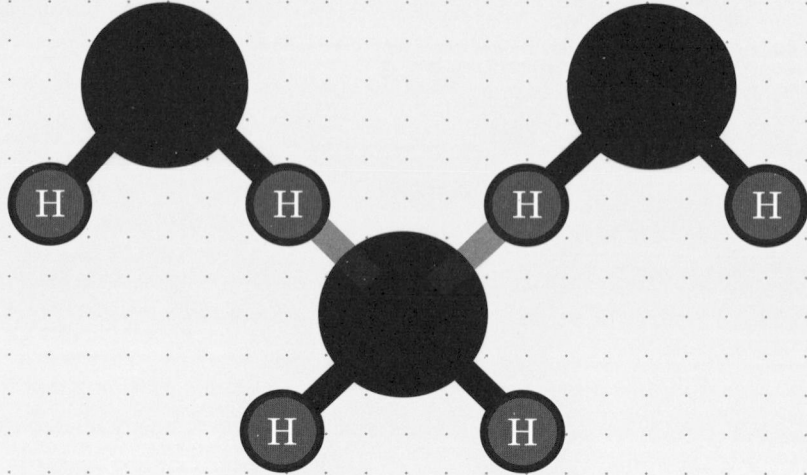

H-bond, you're the bomb!

One of the very first things you learn in high school chemistry is that a molecule of water has two atoms of hydrogen and one of oxygen. These atoms are joined by what we call a 'covalent bond'. But what you may not know is that when you have a heap of water molecules around, the hydrogen atom in one molecule can bond with the oxygen atom in another molecule. We call this bond a hydrogen bond or 'H-bond'.

H-bonds are incredibly cool. Molecules of water bond together and split apart at an amazing pace. Water molecules can change partners billions of times in just a second. At any given time perhaps only 15% of the molecules in a mass of water are actually bonded together but if it wasn't for these H-bonds, the boiling point of water would be more like −90° Celsius rather than the +100° Celsius that it is on Earth.

This would mean water would exist almost exclusively as a gas not a liquid. Life on Earth — yeah thanks for trying, but no way!

H-bonds, I owe you one, big time.

Snip snip!

Three long-haired siblings, Sophia, Salome and Sergio, walk into the hairdressers at 90 Puzzle Place.

They each require a haircut and another procedure. Sophia requires a treatment, Salome a blow dry and Sergio requires a shave.

There are only two hairdressers on duty, Henrietta Haircut and Bob Bob.

They can each do a haircut in 90 minutes (these guys have some serious hair) and the treatment, shave and blow dry each take 30 minutes.

Obviously you need a treatment before a haircut, and a blow dry comes after a snip, but Sergio's shave can happen before or after he gets his locks trimmed.

Further, you cannot cut hair for at least an hour after you've applied the treatment so it can 'sink in'. (So I'm told — a cursory glance at the photo of me on the cover of this book will show you that, when it comes to matters of the hair, I'm purely speculating.)

How should the haircuts, blow dry, treatment and shave be done so as to get our three long-haired siblings out of the shop quickest?

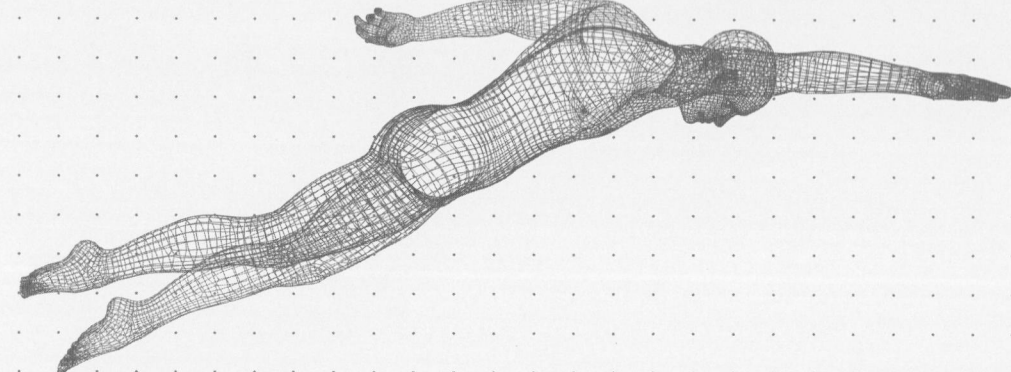

Ossie
Ossie
Ossie

On 9 April 2017, 91-year-old Ossie Doherty teamed up with 3 buddies to set the Australian swimming record for a relay team with a combined age of over 360.

Joined by Ray McGimpsey (90), the whippersnapper youngster Max Van Gelder (87) and led off by the experienced granddaddy of the team John Sheridan (92) they smashed the national age record in a race dedicated to former teammate John-William Steen who died a year earlier, the day before they first planned to chase the mark.

As John Sheridan put it, 'My son reckons that I've always lived dangerously and I probably have. But life's good.'

$$100 = 91 + \frac{\bigcirc\bigcirc\bigcirc\bigcirc}{\bigcirc\bigcirc\bigcirc}$$

More fun on the run

Use the digits 2, 3, 4, 5, 6, 7, and 8 once each to complete the equation. There are three solutions and to reduce your stress levels slightly, I'll let you know that they each have a 2 and 5 on the top and a 6 on the bottom.

Don't forget the 3-digit and 4-digit numbers you're looking for have to fit into the multiplication algorithm. You might like to look back at the answer for 82 to find a divisibility trick you can use again here.

Good luck!

O's and X's iii

We learned way back at 3 that there are essentially 3 patterns for a drawn game of noughts and crosses. Namely:

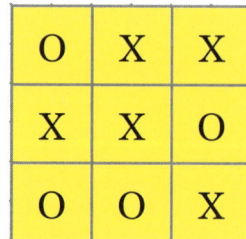

Further, if X goes first there are 91 winning patterns for X and 44 for O. If you've played a bit, it will make sense that it is harder to win if you go second, especially if your opponent has even the faintest idea how to play.

But each winning pattern can be arrived at many different ways. For example, in the result:

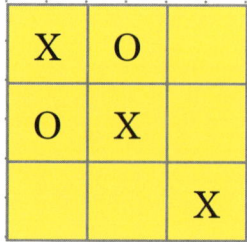

The person playing O could place their moves in those 2 spaces in any order and still end up with the same losing position. Similarly, it doesn't matter which order X played their 3 moves they would still win. There are $2 \times 1 = 2$ possible move orders for O and $3 \times 2 \times 1 = 3! = 6$ move orders for X so the above end position can be arrived at $2 \times 6 = 12$ ways.

You should be able to see that as the end positions become more complex the number of games that could end that way increases. So, for the result:

X	X	O
O	X	O
X	O	X

O can move $4 \times 3 \times 2 \times 1 = 4! = 24$ ways. There are $5 \times 4 \times 3 \times 2 \times 1 = 5! = 120$ different ways to place the Xs, but some of these aren't valid because you can't play the 3 winning spots first or the game would already be over. The cute observation that for every one of the 120 games, one of those 5 Xs has to go last helps us here. As long as the last X is on the winning diagonal we are fine. This will happen in $\frac{3}{5}$s of these 120 games or $\frac{3}{5} \times 120 = 72$ times. So this winning position for X can be reached $72 \times 24 = 1728$ ways!

Across all the possible games of noughts and crosses (this time considering rotations and reflections as different), there are ... maybe you should sit down ... 131,184 ways for X to win; 77,904 for Y and those 3 possible drawn boards can be reached a total of 46,080 possible ways. That makes 255,168 possible games!

Here's the question. How many ways can a game of noughts and crosses end on the 5th move?

Cabbages are 92% water

They also have 18 chromosomes, and when fermented into sauerkraut it is thought that the isothiocyanates produced may limit the growth of cancer cells.
But sauerkraut hasn't always been so popular. During World War I, because the word 'kraut' was slang for German, American sauerkraut makers relabelled their goods as ... wait for it ... 'Liberty Cabbage'.

If you think this sort of behaviour belongs a century ago, in 2003 when France opposed the invasion of Iraq, certain US cafeterias began calling french fries 'liberty fries'.

Chess mess

Blake Black is having a bad day.

He's just played chess against both Cameron Castle and Kyeema King and been well and truly thumped both times.

He turns to his friend Whitney White for some consolation and she is no help at all.

'That's an embarrassing effort, Blake,' says Whitney.

'Bet you couldn't do any better,' replies Blake.

'I bet I could,' she responds, chortling. Hey, with friends like these … 'I've got $92 that says I can play them both and do better than two losses. I'll even play them at the same time.'

Blake accepts the bet.

He does wonder for a second why she wagered an odd amount like $92. It's *almost* as though it was chosen to give us a puzzle for this number in the book, but that's not important — he is *more* than up for the bet.

He knows for a fact that Whitney has practically never played before in her life. Frankly, he's not sure if she even understands all of the rules.

Whitney sets up the two sets, calls over Cameron and Kyeema and plays them both at the same time. She wins the bet … easily.

How?

1AU

Back at 33 we met the beautiful cosmic length measurement the Astronomical Unit or AU.

1AU was originally based on the average distance of the Sun from the Earth over a year but is now formally defined as 149597870.7 kilometres or pretty much 93 million miles.

We used the AU along with G the gravitational constant in calculating a solar mass and I just had to squeeze this amazing little related math moment in here for you.

We now know G to be have a numerical value of 6.67408×10^{-11}, (I'll drop the metres cubed over kilograms seconds squared for here!) but get this. Way back in 1798 the brilliant physicist Henry Cavendish calculated the density of the Earth in his famous Cavendish Experiment. Doing this also gave him a value for the gravitational constant. Cav-Man estimated G as 6.74×10^{-11}. Incredibly, this is only 1% off our current value and he did this without modern measuring machines, computers, or a fidget spinner … truly amazing.

This may explain why the Cavendish Laboratory from back at 29 was named after him.

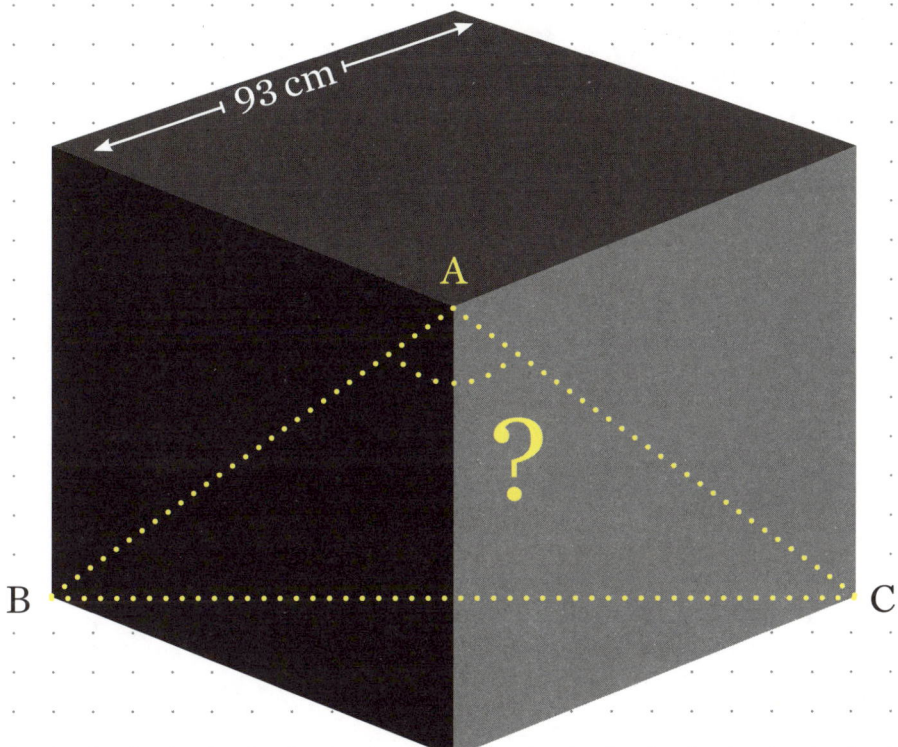

093

As difficult as ABC

We have a cube of sides 93 centimetres and label 3 of the corners of the cube A, B and C as above.

The straight lines joining A, B and C would make a triangle.

Which of the angles in the triangle ABC is the biggest?

And if I triple the lengths of the sides of the cube to now all be a whopping 279 centimetres long, what happens to the sizes of each of the three angles?

Ninety-four is the *only* number ...

... greater than 1, that equals the sum of the squares of the digits of their own square in base 11.

Yep, next time someone happens to mention 94 in conversation, perhaps 'I live at number 94' or 'didn't that song come out in '94?' you can throw in this little fact. Then once everyone has stared at you for a while, you can explain what it means.

We already spoke about base 16, which is used in the language of computers, back at 53. Well, base 11 is like base 16 — sort of.

We count in base 10, so when we have counted off the 10 digits 0, 1, 2, ... 8 and 9 we finish with the digits and move to the tens column. When the tens are full we move to 100, and so on. Right? So when we write 7506 we mean $7 \times 10 \times 10 \times 10 + 5 \times 10 \times 10 + 0 \times 10 + 6$.

Well, in base 11 we would have 11 digits, say 0, 1, 2, ... 8, 9 and A where A is equal to 10 units, and we think of numbers as sums of powers of 11.

So taking 94 we first need to work out 'its own square', which is $94^2 = 8836$.

Now take this square, 8836, in base 11. We break 8836 into powers of 11. Now $11^3 = 1331$ and $6 \times 1331 = 7986$; subtracting 7986 from 8836 leaves 850. $11^2 = 121$ and $7 \times 121 = 847$. This leaves just 3. So 8836 becomes 6703 in base 11.

Now we just need the squares of the digits of this square in base 11 — that is the $6^2 + 7^2 + 0^2 + 3^2$ ... which equals 94.

So there you go!

$$100 = 94 + \frac{\bigcirc\bigcirc\bigcirc\bigcirc}{\bigcirc\bigcirc\bigcirc}$$

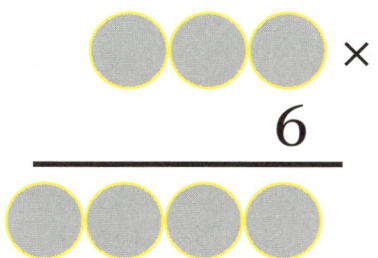

$$\frac{\bigcirc\bigcirc\bigcirc \times}{6} \bigcirc\bigcirc\bigcirc\bigcirc$$

Run for the door!

Using the digits 1, 2, 3, 5, 6, 7 and 8 complete the first fraction above. You've had a bit of practice here back at 81, 82 and 91. So you're looking to fill the second algorithm above.

Think about what digit has to go last in the 3-digit number. In fact, there are only a couple of options for this digit.

When you consider these two cases and try to fill the middle spot in the 3-digit number you quickly eliminate almost all options.

Give it a crack!

The heaviest objects ever to have lived are the giant sequoia trees of the United States

Growing up to 95 metres tall, they are matched in height by other trees around the world, like the amazing Giant Ash Eucalyptus in Tasmania, Australia, but by weighing in at 6000 tons they sure are the bulkiest!

• 15 • 15 • 20 • 10 • 25

• 10 • 15 • 10 • 45 • 15

• 30 • 45 • 50 • 45 • 15

• 20 • 10 • 20 • 20 • 20

• 10 • 15 • 40 • 45 • 50

The spliceman cometh

One last time, with feeling ...

You know the drill by now: split the array of numbers above into sections so that each non-empty region's numbers sum to ... 95.

Mark off these regions using only 3 lines. Each line must be straight and run between two yellow dots. And if a line goes through a black dot, that dot's number has been crossed out and doesn't belong to any region.

Remember that it's the dots that score the points. If a dot is in a region it keeps its number even if the number is crossed out or lies in another region.

One last time, with feeling! Enjoy!

The most tentacles on an octopus ever recorded is a whopping 96

... on a common octopus on display at the Shima Marineland Aquarium in Japan in 1998. This multi-tentacled marvel survived for 5 months but also became the first multi-tentacled octopus known to give birth. It laid a batch of eggs and all the young born from the batch had only two tentacles. Unfortunately they also lived fairly short lives.

You'll note that I've referred to them a 'tentacles' and not 'legs'. That is because, contrary to the popular belief that an octopus has 8 legs, recent research suggests it would be more accurate to say that they have 2 legs and 6 arms or 8 tentacles all told. Claire Little of the Weymouth Sea Life Centre made over 2000 observations of octopuses doing stuff and found while the back 2 tentacles do help them move along the ocean floor, the other 6 are used mainly for feeding.

One final thing about the amazing tentacles of an octopus — these incredibly versatile organs, which can harden like an elbow joint or fold up and vanish for camouflage purposes, actually contain about two-thirds of an octopus's brain. If you count the number of neurons only about one-third of them reside in the blob of grey matter inside the octopus's head.

$$100 = 96 + \frac{\bigcirc\bigcirc\bigcirc\bigcirc}{\bigcirc\bigcirc\bigcirc}$$

How's tricks?

This is the last of our fiendish fractions for the book.

Again you're looking to place the digits we don't use in 96, that is 1, 2, 3, 4, 5, 7 and 8 into the fraction above, using each digit once and once only.

No hints here. Just smash it out and really concentrate to make sure you don't miss any possibilities.

There are 3 answers for this baby, and if you get all 3 you have done exceptionally well.

5	×	4	—	8
—	2	×	2	×
6	+	8	×	7
—	5	—	3	=
3	—	9	=	96

1	+	6	—	1
×	9	—	2	+
6	+	8	×	5
—	6	×	4	=
2	—	7	=	96

3	—	2	+	1
×	6	+	5	×
4	+	9	×	3
+	2	×	1	=
8	—	6	=	96

Number 96

You know the drill ...
You can move up, down, left or right, but you can never visit the same square twice.

Dive on in and find the 3 different paths for each of these grids that trace out the correct equation. And remember: order of operations do apply.

Off you go!

The final chowdown ...

With spring in the air, our 8 diners choose Natu's Nigiri for their November nosh-up, their final group meal of the year.

Now, to make this last question a little bit tougher, I'm not telling you who is sitting opposite Aaron, but you do know the following:

- Each couple (Aaron/Caroline, Barry/Fantine, Danielle/Eamon, Gary/Hayley) sat together
- Each person sat next to both a man and a woman
- Barry didn't want to sit next to, nor face, Gary
- From Eamon's perspective, Caroline was on the left side of the table, and Barry was on the right

Again place Aaron at the top of the table and see if you can recreate the seating arrangement.

Fancy some bonus points? They're all yours if you can look back over all 4 of the restaurant seating questions in this book and find one more little pattern.

And by the way, if you're looking for a super hard challenge... try solving the previous three restaurant seating questions *without* starting off knowing who sat opposite to Aaron!

When a chess player takes on multiple players at the same time it's called a 'simul'

As in, 'simultaneous' play. And just in case you're wondering, the most people that one chess grandmaster has ever played in a simul is 604. Say what? Yes ... 604.

On 8 February, 604 opponents hunkered down over their boards and faced off against Iranian Grandmaster Ehsan Ghaem-Maghami in a marathon 25-hour effort. 'EGM the GM' won an incredible 580 of these games and lost only 8, achieving a final score of 97.35%.

During the simul he is estimated to have walked up to 55 kilometres.

___ × ___ × ___ + ___ ÷ ___ + ___ − ___ = 97

___ × ___ × ___ − ___ ÷ ___ + ___ − ___ = 97

(___ × ___ − ___ × ___) × ___ × ___ + ___ = 97

(___ + ___) × (___ − ___) + ___ × (___ + ___) = 97

(___ + ___ + ___) × (___ × ___ − ___) + ___ = 97

Third time's a charm ...

O kay, so you're getting pretty good at these, yeah?
Well, here's our final bonus round. Same (pretty obvious) rules, but this time use 1, 2, 3, 4, 5, 6 and 8 to complete the equations. Suffice to say we're getting to the pointy end with these punishers.

Get to it!

Ninety-eight is the highest number that will ever be worn by a player in the National Hockey League (NHL)

That is, since the NHL retired the legendary Wayne Gretzky's number 99.

The rules of major league sports tend to forbid 3-digit numbers or higher. They also don't allow decimal places or irrational numbers which sucks because I'd love to see Isaac Heeney from the Sydney Swans swap his 5 for a √5 one week.

Think like a pirate

Have you heard the one about the 5 pirates and their loot? Well, me hearties, settle in for a tale of maths and murder on the high seas.

Returning from a successful lootin' and plunderin', our 5 pirates need to decide how best to divvy up their booty of 100 gold coins.

Each of the pirates has sworn to adhere to the following code, or else he will be made to walk the plank and meet his certain death at the bottom of the briny sea.

The eldest pirate of the 5 must propose a method of dividing the loot. All of the pirates, including the eldest, then vote on whether to accept his terms.

If 50% or more accept, they divvy up the loot and sing ribald sea shanties while scheming their next round of lootin' and plunderin'.

If less than 50% of the pirates agree, the proposer is made to walk the plank, before the *next* oldest pirate proposes a new method, and they all vote again.

Now, remember, these guys are pirates. They're looking out for numero uno, as they say. They're going to make their decisions according to however they think they can get the most money without getting killed. They're also pretty bloodthirsty, and they have no qualms throwing their shipmates overboard. In fact, since they're bloodthirsty pirates, you could probably assume they'd actually *prefer* to throw someone overboard, unless it meant they'd lose money.

Oh, and it's fair to say they don't trust each other. At all. And they certainly won't be making any deals with anyone.

But despite all this, they're pretty logical thinkers, and they're perfectly capable of considering all the possible outcomes.

So, our 5 pirates — A, B, C, D and E (A being the eldest, and E the youngest) — sit down over a bottle of rum and get ready to divvy up their loot.

At the end of it all, pirate A, the oldest, is left holding 98 of the 100 gold coins.

How is that possible?

Adam, you fiend!

Okay, hopefully if you've come this far in the book you've really enjoyed cracking some puzzles.

You may have even got better at it along the way through a combination of some fairly tough ones and my explanations in the answers. So how about we dive into a really tough puzzle to celebrate reaching number 99?

This puzzle appears on the tremendous British maths website www.murderousmaths.co.uk and in the book *Murderous Maths: Professor Fiendish's Diabolical Brain-benders* by Kjartan Poskitt and Philip Reeve (published by Scholastic Children's Books). Professor Fiendish (Kjartan Poskitt) proudly boasts that it is the worst puzzle in the entire book so if you can crack this, you can feel pretty happy with yourself.

Here we go. I throw two special dice. All 12 numbers on the dice are different, but some of the numbers (for example, 19 and 61) look the same when you rotate them 'upside down'. You can pretend there used to be underlines on each face showing which way to read them, but they have since worn off and now only I know the true value of each face.

After the first throw, the dice land like this. I can tell you that the numbers on the bottom of the dice add to make 105.

This is what the dice look like after the second throw. This time the numbers on the bottom of the dice add to make 149.

I throw the dice one more time. Here's the big question ... What do the numbers on the bottom of the dice add to make now ... 100, 127, 159 or 187?

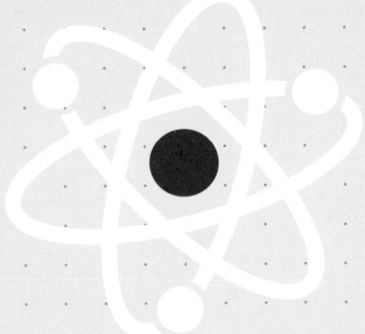

Over 99% of the mass of any atom is in the nucleus ...

... the bunch of protons and neutrons in the atom's 'core'. The remaining 1% of the mass are the electrons whizzing around away from the nucleus.

This goes for the atoms that make up us and the atoms that make up the chair you're sitting on right now. But here's the cool bit: while you may be 'sitting' on the chair, you are not actually touching the chair. Huh?

I don't mean that you have incredible core strength and thighs of steel and are holding a squat just above said chair. I mean that at the micro-scale, no part of the molecules that make up you are actually touching any part of the molecules that make up the chair.

Sure, your butt is 'close' to the chair. But the electrons buzzing around your skin cannot get closer than about 0.000001 centimetre from the chair electrons before they repel each other. If it weren't for this repellent force, the force of gravity would pull you straight through the chair and everything else on the way to the Earth's core.

So why do we 'feel' the touch? That's how our brains interpret the interaction between electrons when they get incredibly close together.

One hundred!

We've spoken in this book about factorials, nifty numbers written like 4! where $4! = 4 \times 3 \times 2 \times 1 = 24$.

Factorials get very big very quickly. Just 62! — the number of ways 62 students can line up to get into a museum — is larger than the number of fundamental particles in the entire universe.

Take a second to think about that ... 62! is easy to understand as $62 \times 61 \times 60 \times 59 \times 58 \times ... \times 3 \times 2 \times 1$, but if you actually crunch it out, 62! = 3146997326 0387937525665312235495076408801228079725823219216316824782110720 0000000000000, a massive 86-digit long beast.

So, consider the number gained by multiplying together the first 100 factorials, namely $1! \times 2! \times 3! \times 4! \times ... \times 99! \times 100!$

I won't get you to work this number out, it would truly break your mind. But, instead, tell me this:

$1! \times 2! \times 3! \times 4! \times ... \times 99! \times 100!$ is not a square number, like $3^2 = 9$ or $7^2 = 49$ are. But if you remove one of the factorials from the product, the remaining product is a perfect square.

Which factorial do you have to remove?

Need a hint? This is really tough and you should feel very good about your calculation abilities if you crack it.

Notice that $1! \times 2! = 1! \times 1! \times 2$, that $3! \times 4! = 3! \times 3! \times 4$, that $5! \times 6! = 5! \times 5! \times 6$ and so on. Rewrite all the terms in the product this way and see how you go. Maybe just try the easier example of the product up to 12! first and see if you can make sense of it. Good luck!

A single bite from an Inland Taipan is thought to have enough venom to kill 100 adult males

One thing I love about Australia is the fear that just mentioning the name of my country can strike in the hearts of people around the world. While some hear 'Australia' and think of beautiful beaches, the Sydney Opera House and the Great Barrier Reef, others quiver in terror at the thought of thousands of different animals that could kill you just by looking at you as you make your way innocently down the street.

The truth is very few people die from encounters with 'nasties' down under. In fact, each year more people are killed by good old-fashioned horses than by all our horrid bees, wasps, snakes, spiders and jellyfish combined. But that doesn't mean we aren't home to some classic critters.

One of my favourites is the Inland Taipan, also known as the Fierce Snake or *Oxyuranus microlepidotus* to any herpetologists out there. Thanks to a combination of it living mainly in remote inland areas, being a pretty chilled out dude and the existence of an antivenin, as far as we know no one has actually ever died from being bitten by an Inland Taipan.

But what a bite it has. Its venom is the strongest of any snake ever measured, by a long way.

Seriously, just leave them be.

Perfectly puzzling pecking poultry

Let's finish our puzzle quest with one more brain-buster from the Mathcounts National Mathematics competition in the United States.

In a barn, 100 chicks sit peacefully in a circle. Suddenly, each chick randomly pecks the chick immediately to its left or its right. Each chick pecks only once, and is not affected by which way its neighbours peck. What is the most likely number of unpecked chicks?

The 2017 competition was won by Luke Robitaille of Texas, a 13-year-old boy who took less than a second to answer this! Now again I stress I don't expect you to get it in less than a second, and little Luke has spent thousands of hours practising this sort of stuff, but have a good long think about it, maybe using some of the logical ability you have picked up throughout this book and see if you can find the answer.

If you're still stuck, here's a hint. Imagine you are one of the chickens in the circle. What are all the possible 'peckings' that could happen to you, including not getting pecked, and what are the odds of each of those 'peckings' happening?

Run these odds out over the 100 chickens and what number of 'no pecks' do you get?

Phew!

Here come the answers to the puzzles in the book.

I don't need to tell you that some of them are really (REALLY!) tough.
So don't be down on yourself if you didn't get all the answers.

For puzzles that you couldn't solve, what I'd love you to do is read
through the answers and then work through the puzzle again,
to see not just *the answer, but the way to* get *the answer.*

Think of it as an aerobic session for your brain!

Answers ...

001

23	6	19	2	15
4	12	25	8	16
10	18	1	14	22
11	24	7	20	3
17	5	13	21	9

By a process of elimination, you should complete the square like this, with the 1 in the very middle square.

Why do the rows and columns all add to 65?

Well, once you've finished the square, all 5 rows contain the numbers from 1 to 25.

And $1 + 2 + 3 + 4 + ... + 22 + 23 + 24 + 25 = 1 + 25 + 2 + 24 + 3 + 23 + ... + 12 + 14 + 13 = 12 \times 26 + 13 = 325$. If 5 equal rows add up to 325, then each row sums to $325/5 = 65$.

001b

Start with all the pages with a 1 in the last position: 1, 11, 21, 31, ... 91, 101, 111, 121, 131, ... 191, 201, 211, ... 291, 301 ... 401 ... 901 ... 991, 1001, 1011, ... 1141 — there are 115 of these pages.

For a 1 in the 'tens' position, we have pages 10, 11, 12, 13, ... 18, 19, 110, 111, 112, 113, ... 119, 210, 211, ... 219, ... 911, 912, ... 919, 1010, 1011, 1012, ... 1019, 1110, 1111, 1112, ... 1119. If you count carefully you will see that are 120 pages like this.

For 1s in the 'hundreds' position, you need the numbers 100, 101, 102, ... 199 and 1100, 1101, ... 1150 which give us 151 pages.

And the number 1 occurs in the 'thousands' for pages 1000, 1001, 1002, ... 1150 — that is 151 pages.

So for pages 1 to 1150 we have 115 + 120 + 151 + 151 = 537 ones.

Apart from getting the correct answer, I hope this question gives you a feel for creating a good 'strategy' for attacking a problem so that you don't miss any important part of the answer.

001c

The large cube will decompose into $3 \times 3 \times 3 = 27$ smaller cubes. The smaller cubes with exactly 1 yellow face will be those that were originally at the centre of each of the larger cube's 6 faces. There will be 6 of them.

002

This problem shows off the amazing power of doubling or what we call 'exponential growth'. You would win the $1,000,000 on ... 21 January. The potential win is $1 (1 January), followed by $2 (2 January), $4 (3 January), $8, $16, $32, $64, $128, $256, $512, $1024, $2048, $4096, $8192, $16,384, $32,768, $65,536, $131,072, $262,144, $524,288 and finally on 21 January, if that 21st head in a row comes up, you walk away with $1,048,576.

A handy thing to note here is that multiplying ten 2s together gives $2^{10} = 1204$ which is very close to

1000. So twenty 2s in a row multiply to give about 1000 × 1000 = 1 million. So you'd expect the odds of 30 heads in a row to be about 1 billion to 1 and it is (1 in 1,073,741,824 to be precise).

002b

This is a famous logic problem with an answer that is beautiful once you see it, but very hard to get first up. Ask twin A, 'What would twin B say if I asked him "is your dad at home"?' Let's say A says, 'He would say, "Yes".' If A is the truth-teller then when B says 'Yes' B would be lying, so you know that Lionel is not at home. If A is the liar, then when he says, 'B would say 'Yes'' A is lying; B would actually say 'No'. Again Lionel is not at home. So If A says 'Yes', Lionel is not at home. By a similar argument, you should be able to convince yourself that if A says 'No', Lionel must be at home. So this single question tells you if Lionel is at home or not.

With the Fibbs siblings, if the black-haired child is telling the truth, then they are a boy. But that means the redhead is a girl and is also telling the truth. This can't happen. So the black-haired child is lying. They must be the girl. This means the redhead is also lying and is the boy. Both Fibbs kids are lying to you!

003

Flip the first switch, then wait a few minutes. Now flip the first switch back to its original position, and then flip switch number 2.

Now, open the door to the basement. Obviously, if the light is on, it is controlled by switch number 2.

If the light is off, it's obviously controlled by switch number 1 or number 3. Go and touch the bulb with your hand. If the bulb is hot, it must have had enough time to warm up — so switch number 1 must control it. But, if the bulb is cold, then switch number 3 — which you didn't touch — controls it. Huzzah!

003b

If he wanted a pair, the worst case scenario would be that the first 3 socks he grabbed were all different. But by grabbing a fourth sock, he must have a pair. So 4 socks will do. If he really wants a pair of polka-dotted socks, the worst case scenario would be if he grabbed every red and every green sock and still only had one polka-dotted sock. This would be 41 socks. By grabbing 42 socks he would have to have a pair of polka-dotted socks. He would also have a fair bit of tidying up to do, wouldn't you, Dave!

003c

There are 3 coins I could sink first, A, B or C. Once I've sunk one, say B, I have two choices of second coin — in this case A or C. Once I sink one of these two, say C, there is only one option (A) that can be sunk last. So there are 3 × 2 × 1 = 6 ways to do this. We write 3 × 2 × 1 as 3!, pronounced 3 factorial, and it's easy to see that 3! = 6. Similarly, if there had been 5 coins there would be 5! = 5 × 4 × 3 × 2 × 1 = 120 ways to nurdle them into the hole.

To make sure you write down all 6 ways, don't just rush in. Construct a logical order to do it. Let's try to write the list alphabetically. So ABC and ACB are the two ways if A is sunk first. Then BAC and BCA if B goes next and finally CAB, CBA.

Bonus points! We need to work out what value of n solves $n!$ = 720. Calculate these yourself if you want to become a maths gun. You should get 2! = 2, 3! = 6, 4! = 24, 5! = 120 and 6! = 720, so I can nurdle 6 coins 720 ways.

003d

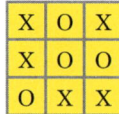

003e

Answer 1) Most people wrongly think, 'The box that had a gold coin can't be the two silvers box — it has to be either the two golds or the "one of each" box. So the odds of the other coin being gold is $1/2$.' Wrong! In fact it's $2/3$! Huh? Think of it this way. Let's say the gold-gold box has coins G1 and G2 and the gold-silver box has coins G3 and S1. You have drawn out a gold coin. It is equally likely to be any of G1, G2 or G3. If it is G3 you will draw out S1 and fail. But if it is G1 you will draw out G2 and if it is G2 you will draw G1. So you have a $2/3$ chance of drawing a gold coin next.

Answer 2) Draw from the box labelled 'gold and rocks'. If you get a gold coin, this box must be gold only. Take that box. But if you draw a rock, this box must be rocks only. The box labelled gold can't be gold either. So the box labelled rocks is in fact the box of gold. Take it and run!

004

It's not possible for all 4 of the statements to be false. If that were the case then statement 4 would be true and we wouldn't have 4 false statements. But no two of the statements can be true because they would contradict each other. So only one statement is true and three are false. The true statement is C.

004b

Answer 1) There are 4 choices for the left-hand stripe of the flag; 4 choices for the middle and 4 choices for the right-hand stripe. So there are 4 × 4 × 4 = 64 possible flags. Note that if our flags had, say 7 stripes, the answer would just be 4 × 4 × 4 × 4 × 4 × 4 × 4 which we write as 4^7 (to save writing all seven 4s).

Answer 2) There are 4 colours that could go on the left-hand stripe of the flag, but once we've chosen a colour, say white, for that stripe there are only 3 colours we could choose for the centre. Once we've chosen the left-hand and centre colours there are only 2 colours left for the right-hand stripe. So there are 4 × 3 × 2 = 24 possible flags. This process where we multiply numbers by descending whole numbers occurs often in counting problems. In particular, if we had 7 colours and 7 stripes there would be 7 × 6 × 5 × 4 × 3 × 2 × 1 possible flags. As you just saw in our nurdling question, we write this as 7! and call it 'seven factorial'. FYI, 7! = 5040.

Answer 3) There are 64 possible flags and 24 do not repeat a colour. All the other flags must repeat a colour in some way, so 64 – 24 = 40 flags involve some repetition of a colour. This answer involves a common strategy in counting problems. It's hard to work out all the different examples of flags with a repeated colour, but clearly they are the same as all the flags that don't qualify for the second group we just counted. So just take those 24 away from the full set of 64 and *booyeah*!

Answer 4) There are 4 choices for the left-hand stripe, but only 3 choices for the centre stripe because we can't repeat the left-hand side. But any of the remaining 3 colours (not the colour we chose for the middle) could then go on the right. So we have 4 × 3 × 3 = 36 possible flags.

Answer 5) There are 24 flags that do not repeat a colour. But if you made all 24 of these flags, every flag in your pile would have a buddy flag that is the same flag in reverse order — Red Green Blue buddies up with Blue Green Red, for example. So the list of 24 flags halves to a list of 12 placemats with no repetition of colour. Again, this is a common occurrence in counting problems. Within a larger set, in this case 'all flags with no repeated colour,' we group together objects that we now consider to be effectively the same and divide the size of the set down, in this case $^{24}/_2$ = 12.

004c

We work backwards. Since the final traveller must have eaten $1/_3$ of the nuggets on the plate, there are $2/_3$ of that amount left. Four is $2/_3$ of 6, so, there must have been 6 nuggets when she began. Similarly, 6 is $2/_3$ of 9 and $9/(2/_3)$ = 13.5 nuggets. So, adding the extra half nugget that the room service guy ate, the manager had prepared 14 nuggets.

005

If Peter was told the product was 5 he would know the cards have to be 1 and 5. If he was told the product was 12, it could only be 3 and 4. There is only one product that could leave him unsure and that is 4, because it could be 1 and 4 or 2 and 2.

So Sally knows it has to be 1 and 4 or 2 and 2. And she has already been told the sum. If she tells Peter the sum is larger than the product, Peter knows she must have been told the sum was 5. So the cards are 1 and 4.

005b

006

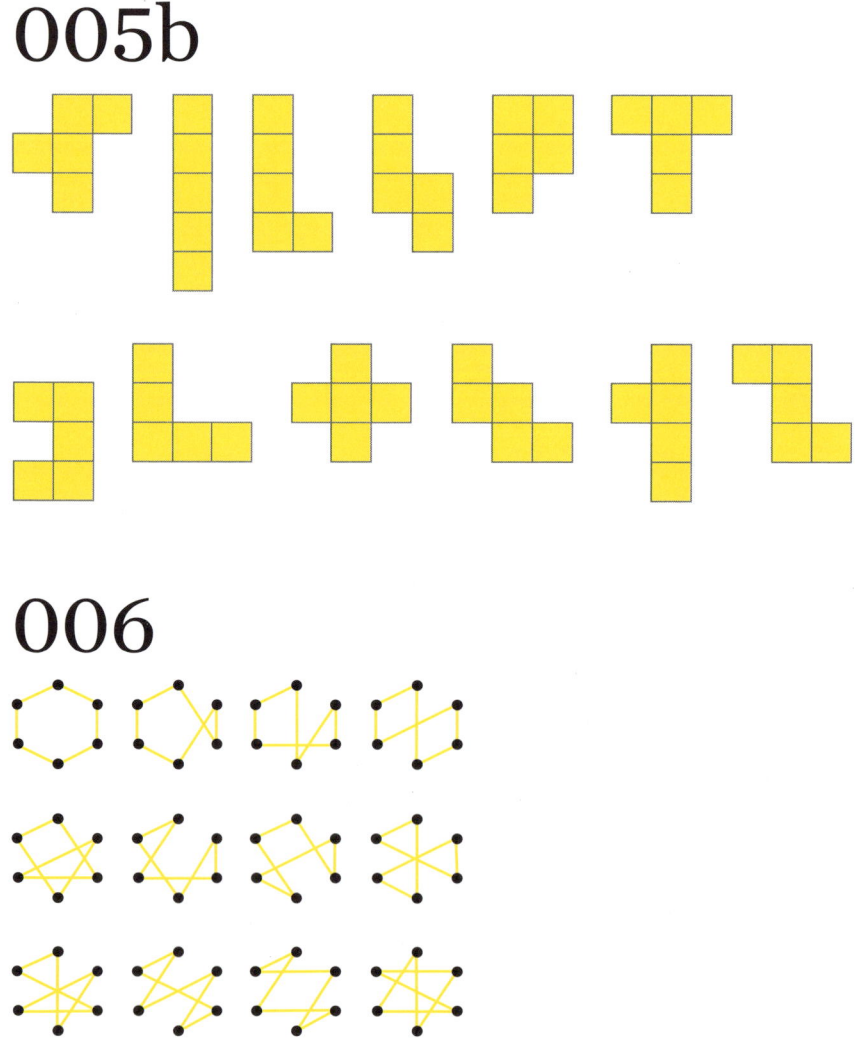

006b

Name the 6 people A, B, C, D, E and F. If you just start randomly shaking hands and trying to keep count you will never get this, so let's arrange it logically. If A shakes hands with B, C, D, E and F that gives us 5 handshakes. A can now sit down. Now B has already shaken with A but needs to shake hands with C, D, E and F then, after these 4 handshakes, she can take her place at the table. Similarly, C has already shaken with A and B but not D, E or F so has three shakes to make. D shakes hands with E and F and then once E shakes F's hand we are done. This makes 5 + 4 + 3 + 2 + 1 = 15 handshakes. We call 15 the 5th 'triangular number'.

In general if n people shake hands, there will be the $(n-1)$th triangular number of handshakes.

006c

In the first example we simply add the dice so the final roll scores 12. In the second game we multiply the dice $1 \times 5 \times 6 = 30$, $2 \times 3 \times 3 = 18$ and $5 \times 5 \times 5 = 125$, so 4, 4, 4 scores 64 beating 2, 5, 6 which scores only 60. And the third one was a bit nasty. Count the number of 'middle dots' that occur on the faces. By this I mean the dot that is painted to signify a 1. There are middle dots on 1, 3 and 5s but not 2, 4 and 6s. So for example, 5 5 4 contains 2 middle dots. So 6 6 1 would score you 1.

007

You should be able to see that it is impossible to score 11. You can score 12 as 4 laogs which are 3 points each. You can score 13 as a yrt and 2 laogs and you can score 14 as 2 yrts. But once you have 3 consecutive scores that you can achieve, namely 12, 13 and 14, you can get every greater score just by adding as many laogs as you need to 12, 13 or 14.

In the second example, a yrt is worth 11 and the highest you cannot score is 39. If a laog is worth 1 you can obviously get 39 laogs. If a laog is worth 2 you could score 3 yrts and 3 laogs and get 39. If a laog was worth 3 points then 13 of them will get you 39 points. Continue this logic and you'll see a laog must be worth 5.

007b

Let's give the bottles a value in litres. Say the small bottle is 1 litre. The large bottle must be 2 litres, then. Do you see where I'm going with this? Each man should get 7 litres of wine between 4 small and 4 large bottles ... so you should be able to deduce that 'Brother John gave the first man three large bottles and one small bottleful of wine, and one large and three small empty bottles. To each of the other two men he gave two large and three small bottles of wine, and two large and one small empty bottles. Each of the three then receives the same quantity of wine, and the same number of each size of bottle.' Nice one!

008

8! = 8 × 7 × 6 × 5 × 4 × 3 × 2 × 1 = 40,320 so we need to find 4 distinct factorials that add up to 45,369 – 40,320 = 5049. 6! is only 720 so clearly we need 7! = 5040. We still need 5049 – 5040 = 9 from 3 other factorials which are obviously 1! + 2! + 3!. So 213^2 = 1! + 2! + 3! + 7! + 8!

008b

21 (values). You can crack this with hard slog and drawing up a table of results. The least you can have is 1c and the most obviously 50c. Write down the numbers 1, 2, 3, …, 49, 50 and go through number by number circling them if you can make them out of exactly 5 coins, and crossing them off if you can't. So 23 = 10 + 10 + 1 + 1 + 1 gets a circle, but 37 needs at least 6 coins (10 + 10 + 10 + 5 + 1 + 1) so gets a cross and so does 11 because we can get it as 10 + 1 or 5 + 5 + 1 but not with exactly 5 coins.

A more logical way is to think through how many 1c coins are in your final set. If there are 5 ones we have just one result (5). If there are 4 ones we have 2 results (9 = 4 + 5 and 14 = 4 + 10). If there are 3 ones we have 3 results (13 = 3 + 5 + 5, 18 = 3 + 5 + 10 and 23 = 3 + 10 + 10). Continuing this, 2 ones gives us 4 results, 1 one gives us 5 results and no 1c pieces give us 6 different results. In total, this gives us 1 + 2 + 3 + 4 = 5 = 6 = 21 possible totals.

You can get the values 5, 9, 13, 14, 17, 18, 21, 22, 23, 25, 26, 27, 30, 31, 32, 35, 36,40, 41, 45 and 50. The longest run you can get is length 3: 21, 22, 23; 25, 26, 27 and 30,31,32. The longest run you cannot achieve is length 4: 46, 47, 48, 49 and 1, 2, 3, 4.

But how did Chad get his answer so fast? He probably realised the question was really asking how many ways you can choose 2 things from a list of 7 things. Huh?

Imagine after getting a handful of 5 coins, you laid them out in order from smallest to largest, for example, 1c, 1c, 5c, 10c, 10c. If for some reason you couldn't tell the difference between the coins, you could stick two lines in that list to separate it into 1s, 5s, and 10s, making our example: ??|?|??

Any list of 5 question marks and 2 dividers is the same as a set of 5 coins broken into 1s, 5s and 10s. So this question that sounds like it's about coins, to a super trained puzzle solving mind like young Chad's is really a question about 'How many ways can you put 2 dividing lines into 7 possible places?' Note it's 7 places because the lines can go at the very beginning, very end or in between any of the 5 question marks.

Well, there are 7 places the first divider can take, and 6 places remain for the second divider after that. That would give 7 × 6 = 42 possibilities, but we've actually counted every case twice, because choosing the 1st position and then the 2nd is the same as choosing the 2nd position and then the 1st. Divide our answer by 2 and Bob's your uncle: there are $^{42}/_2$=21 ways you can place the dividers which corresponds with the 21 different values we just crunched out painfully above.

Bonus points if you realise that this step of dividing by 2 is exactly the same thing we did in the question about placemats back at 004. I told you it was a powerful technique!

This is a great example of how beautifully written maths problems can actually be asking something completely different to what they seem to be at first glance.

009

6	1	8
7	5	3
2	9	4

We use a similar logic here to what we did for the 5 x 5 magic square back at 001. In this case, $1 + 2 + 3 + ... + 7 + 8 + 9 = 45$ so if all three rows are equal they have to add up to 15. And the 5 has to go in the middle. If you put the 6 in the middle, the 9 couldn't go anywhere. Why? Well, if you put a 4 in the middle, then wherever you put the 1 you need a 10 to make 15. So 5 has to go in the middle. Now just try different combinations and you'll get something like this.

Any other result you might have gotten is just the same grid rotated or reflected.

You could also determine the position of the 5 like this. If you add the four lines through the middle together, you'd get 15 four times, or $4 \times 15 = 60$. But that's the same as adding all the numbers in the square (which is 45), plus the middle number 3 more times. So 3 times the middle number is $60 - 45 = 15$, which means the middle number has to be 5. Neat!

010

If you try 3 keys on the first lock, you will know that the fourth key is a match. So, you only need a maximum of 3 attempts. Similarly, the second lock needs a maximum of 2 tries, and the third only needs one. Once you've matched the first 3 locks to their keys, the remaining key obviously goes in the last lock. So there are, at most $3 + 2 + 1 = 6$ tries needed to match all 4 locks to their key.

011

Yep, this is the answer to the world's hardest Sudoku. Please, don't even pretend that you got it!

8	1	2	7	5	3	6	4	9
9	4	3	6	8	2	1	7	5
6	7	5	4	9	1	2	8	3
1	5	4	2	3	7	8	9	6
3	6	9	8	4	5	7	2	1
2	8	7	1	6	9	5	3	4
5	2	1	9	7	4	3	6	8
4	3	8	5	2	6	9	1	7
7	9	6	3	1	8	4	5	2

012

1	2	3	4
5 E	R	D	A
6 E	Y	E	D

012b

'1234 is a multiple of 4' is false.

This question involves a great skill in mathematics: knowing when a number is divisible by a smaller number. We call these 'divisibility tests'. The test for divisibility by 2 is simple: does the number end in 0, 2, 4, 6, or 8? So yes, 12 is a multiple of 2. The test for divisibility by 3 is really cool — if the sum of the digits of the number is divisible by 3, then the number is divisible by 3. So 1 + 2 + 3 = 6 which is divisible by 3, thus 123 is divisible by 3. Now 100 is divisible by 4, as is 200, 300, and any amount of 100s, so for a number larger than 100 to be divisible by 4 we just need the number made up of the last two digits to be divisible by 4. In this case, the last two digits of 1234 are 34 but 34 is not divisible by 4 so neither is 1234. Numbers that are divisible by 5 end in 5 or 0, so 12345 is divisible by 5. If a number is divisible by 6 it must be divisible by 2 AND 3 so applying both those divisibility tests we see 123456 is even and 1 + 2 + 3 + 4 + 5 + 6 = 21 = 3 × 7, so 123456 is divisible by 6.

'1234 is a multiple of 4' is the only false statement.

012c

Option C. The Zigzag lace uses the least shoelace followed by option B and option A, the type they often do in shoestores to save time actually needs the most shoelace to complete.

In case you're wondering how far each lace travels to complete the lacings, if you call the vertical distance between the two rows of holes 1 unit, then using Pythagoreas's theorem (remember in a right angled triangle with shorter sides length A and B and longer side C we have $A^2 + B^2 = C^2$) you can see that in order the lengths are:

(A) $5 + 5\sqrt{2} + \sqrt{26} = 17.170087...$

(B) $5 + 2\sqrt{2} + 4\sqrt{5} = 16.772699...$ and

(C) $1 + 10\sqrt{2} = 15.142135...$

013

Write the numbers 1 to 13 in a circle. Start counting from 1 and you should remove the numbers 5, 10, 2, then 8. Note you remove 8 here, not 7 because 5 is already gone onto the prison ship and won't be counting again. Eventually you will have the numbers 3, 6, 7, 11 and 12 as the survivors. So you don't want to start counting. To save you and your mate you want to be in position 12 or position 7. That means you want the person 2 spaces to your left, or 6 to your right, to start the counting. You and your friend will survive.

013b

For the numbers $a = 3$, $b = 5$, $c = 8$ and $d = 13$ we get $c^2 - b^2 = 8^2 - 5^2 = 64 - 25 = 39$ and $a \times d = 3 \times 13 = 39$. If you do the other calculations correctly you should get both sides equalling 105, 272 and finally 715.

014

Yep, good ol' Bach was reputed to like 14 because you can 'sum' his surname as: B + A + C + H = 2 + 1 + 3 + 8 = 14.

014b

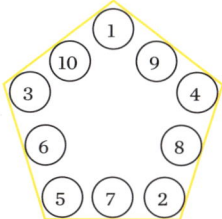

014c

Grab your ruler and prepare to be amazed. The two black circles are actually exactly the same size! It's an optical illusion.

015

There is no extra dollar. There obviously cannot be. So what's happening here? This type of problem is called a 'misdirection' because, like a magician getting you to look one way while they perform a sleight of hand somewhere else, this question gets you to put together things that don't belong to get a nonsensical answer.

Basil and Bonita originally had $30. They paid $25 to the restaurant, the waiter has $3 and they have $2 left over. The $3 goes with the $25 to make the $28 they paid. To then add the $3 again to the $28 is double counting and makes no sense. If you can't quite see this, take 30 buttons or coins, pretend they are all worth $1 each and act it out. You'll soon see the sleight of hand.

016

There are nine 2×2 squares, four 3×3 squares and one 4×4 square. So there are $16 + 9 + 4 + 1 = 30$ squares in our 4×4 grid. Note that $16 = 4^2$, $9 = 3^2$ and $4 = 2^2$. You can extend this pattern and immediately state that the 10×10 grid has $10^2 + 9^2 + 8^2 + ... + 3^2 + 2^2 + 1^1 = 385$ squares within it! Handy, hey?

016b

The increase from 16 to 48 is an increase of 32. So if 16 swans are the usual 100% of the cast, the increase is an increase of 200%. In a sense, it's all about how you word the question. If you drop the word 'increase', then you can actually say that 48 is 300% of 16.

017

$F3 = 2^8 + 1 = 257$ and $F4 = 2^{16} + 1 = 65\,537$.

017b

$7 - 9 \div 9 + 6 + 5 = 17$
$5 - 9 + 5 \times 4 + 1 = 17$
$9 \times 8 \div 4 - 7 \div 7 = 17$
$9 \times 6 - 9 \times 5 + 8 = 17$
$7 + 6 \div 9 \times 5 \times 3 = 17$

These sums give you the highest possible scores. Lower-scoring answers might have put $a + b$ or $a \times b$ where I have $b + a$ or $b \times a$.

018

Amy and Ben cross (2) Amy returns (1); Delia and Cassius cross (10) Ben returns (2); Amy and Ben cross (2). This takes only 2 + 1 + 10 + 2 + 2 = 17 minutes.

You save 2 minutes since even though Ben has to make an extra 2 crossings compared to our 19-minute example, you save 5 minutes by sneaking Cassius across with Delia.

This is a great example of how some puzzles can only be solved when you abandon an assumption you're making about the answer. Tricky puzzles often involve abandoning an assumption that seems so obvious to you, you don't even realise you could do the problem a different way. Serious props if you got this out by yourself!

019

I've written the numbers down alphabetically as they are spelled.

That is eight, eighteen, eleven, fifteen, five, four, fourteen, nine, and so on.

To complete the list 2, 3, 10, 12, 13 would be written ten, thirteen, three, twelve, two or 10, 13, 3, 12, 2.

019b

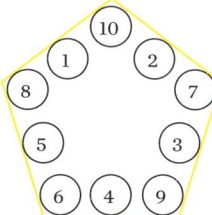

Following the same logic as the 14 puzzle we see that adding all the circles once and the corners a second time must give us a total of 5 × 19 = 95. Now 1 + 2 + ... + 9 + 10 = 55 so the corners have to give us another 40 so we hit 95.

The only way 5 numbers from 1 to 10 can add up to 40 is if they are 6, 7, 8, 9 and 10. So these go in the corners. The 1 must go in between the 10 and the 8. The 5 must go between the 6 and the 8. The 9 can't go next to the 10. You should reasonably quickly get the only solution shown here.

020

R	O	U	F	O
U	F	R	O	U
U	O	F	F	R
R	U	R	O	F

R	O	U	F	O
U	F	R	O	U
U	O	F	F	R
R	U	R	O	F

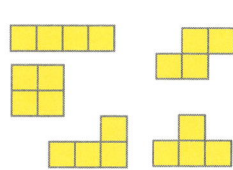

Like I said, there are more. If you get one, feel free to take a photo and tweet it, including my handle @adambspencer or slap it on Instagram and tag me (adam_spencer1).

020b

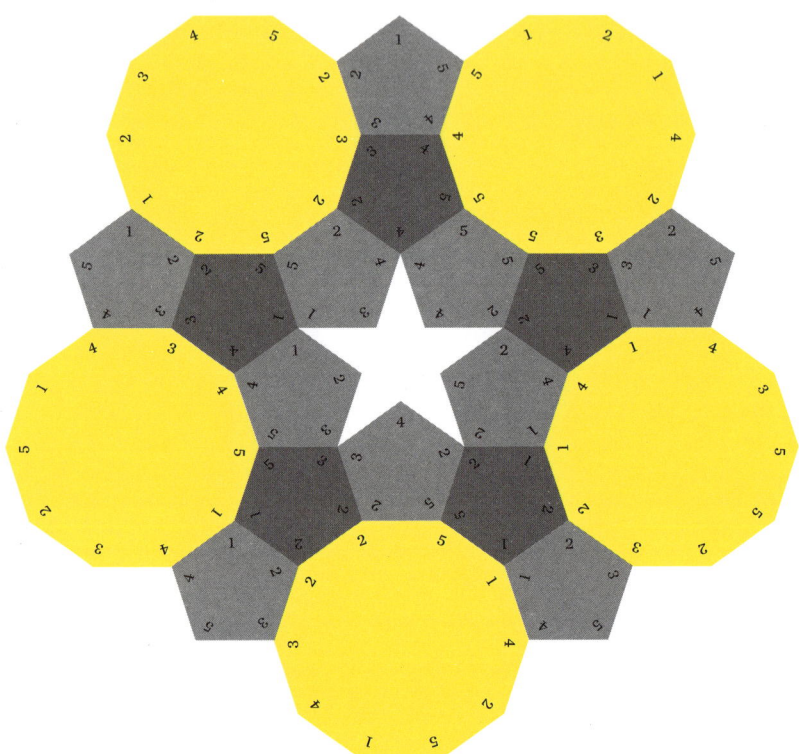

021

The most common symbols in our code and the number of times they occur are † (20), n (14), 8 and × (12), 6 (11), d and 5 (8) and z and ! (7 times each). The most common letters in English change depending on which list you're looking at, but E is always first, and A R I O T N S seem to be the next bunch. This means † could very well be E. When I first saw this puzzle the word †dx††w looked like 'esteem' to me, and I went from there. If that's the case then 8xd occurs twice as does 8x, suggesting 8 = I giving ITS and IT. x6 is possibly TO giving 6 = O. The word 8nd)Pxd is now I_S_ _ TS and the only word I could see was INSULTS. A bit more plugging away and we come up with the quote:

INJURIES MAY BE ATONED FOR AND FORGIVEN BUT INSULTS ADMIT OF NO COM-
PENSATION;
THEY DEGRADE THE MIND IN ITS OWN ESTEEM AND FORCE IT TO RECOVER ITS
LEVEL BY REVENGE.
JUNIUS

021b

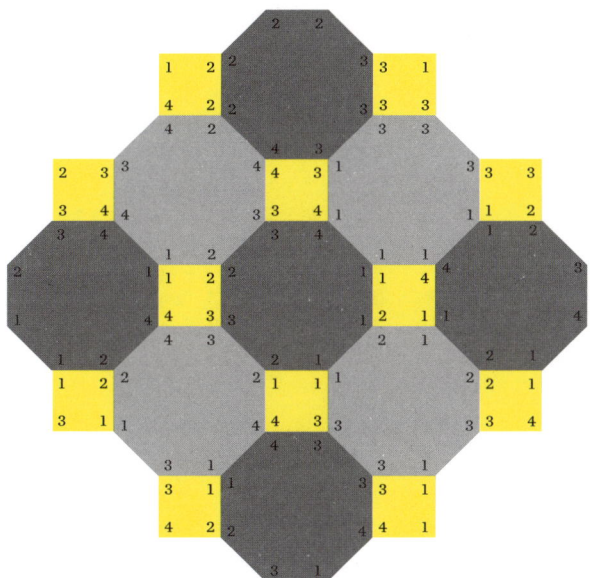

022

8 has 22 partitions, because 8 can be written as: $8 = 7 + 1 = 6 + 2 = 6 + 1 + 1 = 5 + 3 = 5 + 2 + 1 = 5 + 1 + 1 + 1 = 4 + 4 = 4 + 3 + 1 = 4 + 2 + 2 = 4 + 2 + 1 + 1 = 4 + 1 + 1 + 1 + 1 = 3 + 3 + 2 = 3 + 3 + 1 + 1 = 3 + 2 + 2 + 1 = 3 + 2 + 1 + 1 + 1 = 3 + 1 + 1 + 1 + 1 + 1 = 2 + 2 + 2 + 2 = 2 + 2 + 2 + 1 + 1 = 2 + 2 + 1 + 1 + 1 + 1 = 2 + 1 + 1 + 1 + 1 + 1 + 1 = 1 + 1 + 1 + 1 + 1 + 1 + 1 + 1$, and 14 has 22 strict partitions, because 14 can be written as $14 = 13 + 1 = 12 + 2 = 11 + 3 = 11 + 2 + 1 = 10 + 4 = 10 + 3 + 1 = 9 + 5 = 9 + 4 + 1 = 9 + 3 + 2 = 8 + 6 = 8 + 5 + 1 = 8 + 4 + 2 = 8 + 3 + 2 + 1 = 7 + 6 + 1 = 7 + 5 + 2 = 7 + 4 + 3 = 7 + 4 + 2 + 1 = 6 + 5 + 3 = 6 + 5 + 2 + 1 = 6 + 4 + 3 + 1 = 5 + 4 + 3 + 2$.

023

1	5	3	0	4	6
1	4	0	2	3	5
1	2	2	0	1	5
4	5	5	2	1	7
3	2	2	3	5	9
9	9	3	1	9	14

3	5	4	3	3	3
4	4	3	5	5	4
4	3	5	4	3	5
4	3	4	4	4	3
4	5	3	4	4	3
4	3	4	3	4	5

024

This is an awesome example of maths that seems to mess with your mind. For each individual roll, the probability that A rolls a higher number than B, the probability that B rolls higher than C, and the probability that C rolls higher than A are all 5/9, so over time, no one is favourite to win.

We call these types of dice 'nontransitive'. Normally in a contest if A should beat B and B should beat C then we can assume that A should beat C. This relationship is called transitivity and is similar to arithmetic where we know for numbers a, b and c if $a > b$ and $b > c$ then $a > c$. But these dice do not display transitivity! If we use the '>' sign to mean 'will beat over time' we have A > B, B > C but C > A!

024b

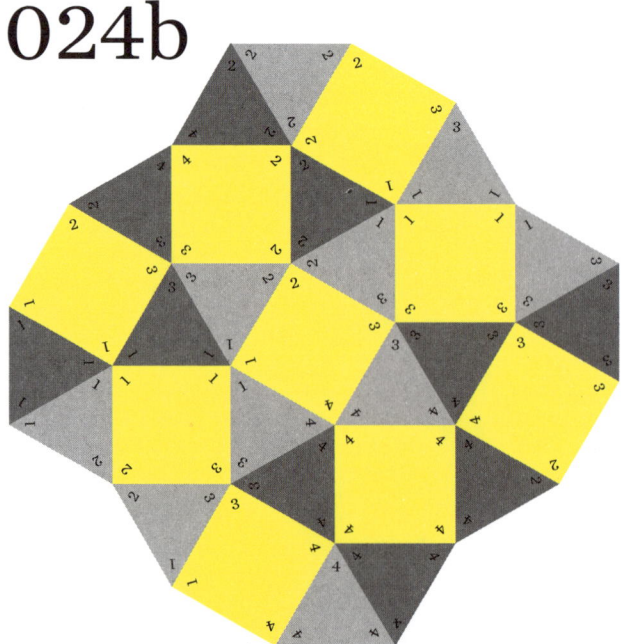

025

The first grid contains the numbers THREE, SEVEN, EIGHT, FORTY and SIXTY. With the second grid, the hint I gave you should have helped you find the countries CHINA, INDIA, JAPAN, EGYPT and ITALY. But with no hint, you probably struggled to find the five well-known fruits! PEACH, APPLE, MANGO, LEMON, GRAPE.

I love this puzzle because none of the answers themselves are obscure words, but the difference in the strength of the hint provided (or not) effects the difficulty of the question so much.

025b

You should have discovered the chain of results A beats B, B beats C, C beats D, D beats E and E beats A. But here's the cool bit. There is a second chain. A beats C beats E beats B beats D beats A.

And here is the *super-dooper* cool bit! If you roll two of each die and add the totals, A still beats B and so on for the first chain. But compare A and C. When adding two dice can you see that C now beats A? This is because two Cs always gives you a score of 10 and A can get 14s but this won't happen as often as A gets 9s and 4s. Keep going along the second chain and you'll see it completely reverses for the double dice game. WOW!

A note to any schoolteachers or parents of youngsters reading this book. Go to bit.ly/2v9PFcP and make up these dice on blank cubes of wood or cardboard and get kids to play the games I've explained above. A great way to expand their little maths minds!

026

The odds of 10 straight heads is, as I told you, 1 in 1024 because $2^{10} = 1024$ so $(1/2)^{10} = 1/1024$. Like we did back in our little coin tossing game at 002, let's approximate this to about 1 in 1000.

Like I said back then, this approximation is handy when estimating large powers of 2. So the odds of 20 heads in a row is about $1/1000 \times 1/1000$ or 1 in a million. After you've tossed 20 straight heads, tossing another 6 heads is a $1/64$ chance ($2^6 = 64$) so a good guess for the odds of 26 straight heads should be 'a little worse than 1 in 64 million'. Indeed the answer is 1 in 67,108,864. Note I've used a coin here because on a casino roulette wheel there is also a 0 and perhaps a 00 slot which changes the odds significantly.

In fact compared to tossing 26 heads, the on a roulette wheel with a single 0 but no 00 the odds of 26 straight reds doubles to 1 in 136,823,184 and with a 00 it balloons to 1 in 273,708,560.

026b

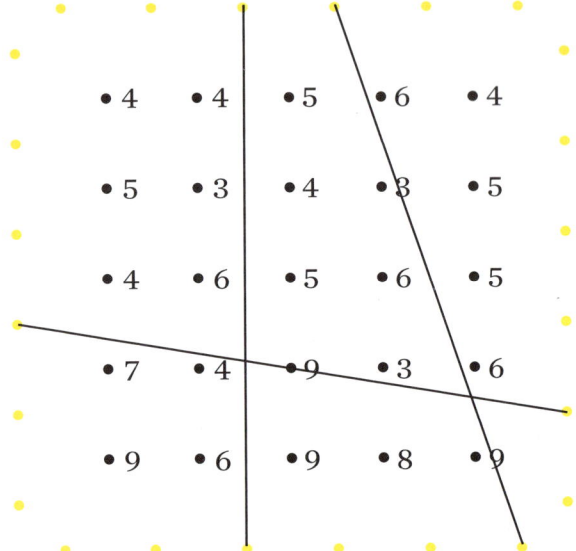

027

You should be able to see that any corner cube will have 3 yellow faces. And the 'centre cube' on each face will have one yellow face. There are 8 corner cubes and 6 centre cubes, so 14 of the smaller cubes will have an odd number of yellow faces.

You should be able to see that there will be 12 smaller cubes on the original cube's edges with two yellow faces. This gets us to 14 + 12 = 26 cubes. The 27th smaller cube is the one in the very centre which is the only cube to have no painted faces at all.

027b

Answer 1) You remove one cube from each face and the centre cube making 7 cubes removed. So 20 (27 – 7) cubes remain.

Answer 2) The Menger sponge has zero volume and infinite surface area.

Okay, go and lie down for a while. Once you've recovered, you You should be able to see that when you remove the first set of cubes from the big cube that we started with, the new, first-stage Menger sponge has volume 20/27 of the original cube. But then you hollow out every remaining smaller cube in the sponge, so the volume of the second-stage Menger sponge is 20/27 of the first stage sponge; or 20/27 × 20/27 of the original sponge. Can you see that the

3rd-stage sponge has volume $(20/27)^3$ of the original? And so on. After 10 stages the volume is $(20/27)^{10}$ which is already reduced to 0.05 or so of the original volume. As this process goes on further and further you can see that this volume gets closer and closer to 0. So if we could apply the hollowing process forever, the volume would in fact reach 0.

Calculating the surface area change per step is a bit trickier, but you can definitely see that the surface area has always increased after applying a hollowing-out step ... because even though you have removed up to 6 smaller squares from each cube's face, the 'tunnels' you've created contribute far more than 6 smaller squares to the new surface area. If you really want to have a hard-core maths nerd go at it, see if you can show the surface area after hollowing out n times is $2 \times (20/9)^n + 4 \times (8/9)^n$. And yep, that number gets infinitely large as n gets bigger and bigger.

028

5	+	6	×	4
+	5	×	5	×
8	+	8	+	7
×	7	−	8	=
8	×	9	=	28

5	+	6	×	4
+	5	×	5	×
8	+	8	+	7
×	7	−	8	=
8	×	9	=	28

5	+	6	×	4
+	5	×	5	×
8	+	8	+	7
×	7	−	8	=
8	×	9	=	28

5	+	8	+	2
×	5	×	6	+
6	×	2	−	7
×	7	+	6	=
6	×	4	=	28

5	+	8	+	2
×	5	×	6	+
6	×	2	−	7
×	7	+	6	=
6	×	4	=	28

5	+	8	+	2
×	5	×	6	+
6	×	2	−	7
×	7	+	6	=
6	×	4	=	28

9	×	5	×	7
−	9	×	5	×
8	+	2	×	3
−	8	−	4	=
6	×	5	=	28

9	×	5	×	7
−	9	×	5	×
8	+	2	×	3
−	8	−	4	=
6	×	5	=	28

9	×	5	×	7
−	9	×	5	×
8	+	2	×	3
−	8	−	4	=
6	×	5	=	28

029

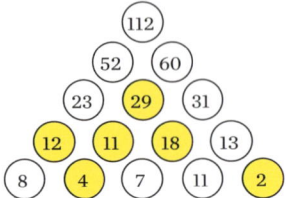

029b

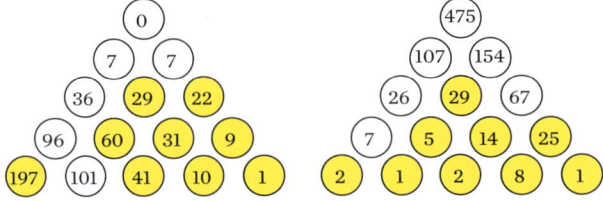

The rules are as follows: for the first, in each triangle of circles, the top circle is the bottom-left circle minus the bottom-right circle. In the second, in each triangle of circles, the top circle is 3 times the bottom-left circle plus the bottom-right circle.

029c

1	0	0	0	0	1
0	1	1	1	1	2
0	1	1	1	1	2
0	0	0	1	1	2
0	0	0	1	1	2
0	2	2	1	1	1

1	0	0	1	1	1
2	1	1	0	0	2
2	1	1	0	0	2
2	1	1	0	0	0
2	1	1	0	0	0
1	1	1	1	1	0

0	0	0	1	1	2
2	0	0	0	0	1
2	0	0	0	0	1
1	1	1	1	1	0
1	1	1	1	1	0
2	2	2	1	1	1

The total sum of each jigsaw puzzle is — you guessed it — 29!

030

No dollar goes missing. Sergio sells his apples for ⅔ the cost of Maria. So if Sergio gives more apples than Maria in a ratio of 3:2 things equal out. Say he gives 30 apples and she gives 20 then they would each have made $10 separately and they make $20 between them. But if the number of apples they supply are equal they lose $1 in total for every 60 apples they hand over.

030b

1	3	5	7	9	11
23	21	19	17	15	13
25	27	29	31	33	35
47	45	43	41	39	37
49	51	53	57	59	61

1	3	4	10	11	20
2	5	9	12	19	21
6	8	13	18	22	27
7	14	17	23	26	28
15	16	24	25	29	30

113	41	43	47	53	59
109	37	5	7	11	61
107	31	3	2	13	67
103	29	23	19	17	71
101	97	89	83	79	73

For the first grid, the path follows odds, winding left or right. In the second, the numbers wind diagonally, increasing consecutively. And in the third, we follow the primes in an outward spiral.

031

As I suggested, let's come to the answer by thinking about the question logically, without any calculations.

Imagine at the end of the month of May you write down a list of the entire list of tosses you made across the month. It will look something like HTTHTHTHHTTHTTTTTHTHTTTHTH-THTHHHTHTTHTHTTTTTH ... and will stop exactly when you have thrown your 31st H. On average how long would you expect that list of tosses to be? Can you see on average it has to be 62 tosses? Otherwise you're suggesting the total list of results is more likely to come up H or T and this can't happen.

Another way to ask the question is, if I tossed a coin 62 times, how many heads would I expect to get? That's right, you have to expect 31 heads or you've got a biased coin.

So you should be able to see across the 31 days of May I should toss, on average 62 times. There's no guarantee for an individual doing this routine that they will toss the coin exactly 62 times. But if, say, 20 people did it, the average number of tosses they took should be pretty close to 62.

031b

There are 31 shapes ... including the white hexagons!

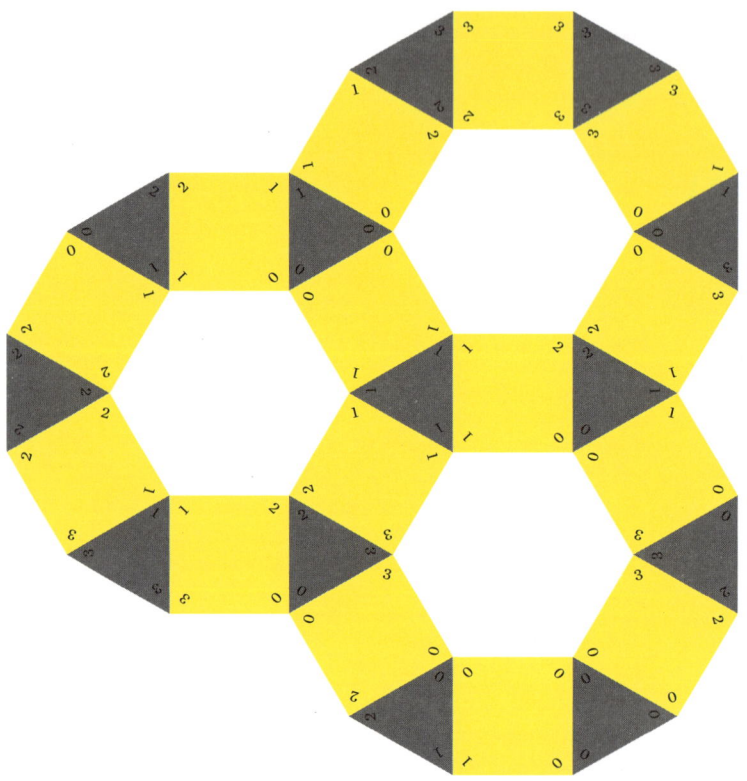

032

$6 × 5 × 4 ÷ 8 + 9 + 7 + 1 = 32$
$9 × 7 − 6 × 5 − 8 ÷ 4 + 1 = 32$
$9 × 8 − 5 × 4 × (7 − 6 + 1) = 32$
$((8 × 5 + 9) ÷ 7 − 1) × 6 − 4 = 32$
$((9 + 8) × (5 + 4) + 7) ÷ (6 − 1) = 32$

032b

There are many solutions to this problem online. Just search 'four fours problem'. But if you did get either $31 = \frac{(4! + 4)}{4} + 4!$ or $31 = \frac{((4 + \sqrt{4})! + 4!)}{4!}$ ... good job!

033

There is a (comparatively) simple solution to this question that many of you would have heard of when at school. If you start at the North Pole and walk 33 kilometres south, 33 kilometres west, then 33 kilometres north, you'll end up back at the pole. Well done if you got that!

But in fact there is a much more complex answer as well. Find the southern latitude on the globe whose full length is 33 kilometres. Start 33 kilometres above it. Go south 33 kilometres. Now go 33 kilometres west and ... you've essentially circumnavigated that entire latitude! So, going north 33 kilometres will take you back to ... you guessed it ... right where you started.

You can see that the latitude that is 11 kilometres in length will also work (since you'll just circumnavigate it 3 times instead of once), and in fact any latitude whose length is $33/n$ for any whole number n (you go south, then around the latitude n times before going north to your starting point).

This problem is a great introduction to understanding how a sphere and a flat surface behave differently as geometric objects. In fact, this very question is often one of the first questions you'll encounter in a 2nd or 3rd year university mathematics course with a scary sounding name like 'introduction to differential geometry'.

It's also said to be one of tech guru Elon Musk's favourite questions in a job interview. So make sure you read it again before applying to Tesla.

034

The concatenation 12 won't give us enough to get near 100 so we must be joining two numbers from 5 through to 9. You can see that adding 89 to 34 gets you to 123 and subtracting all the remaining terms would leave you above 100. So the concatenation we need is either 56, 67 or 78. But 34 is even so we need an even sum. Can you see that taking 56 leaves us needing an even number, 10, out of the digits 1, 2, 7, 8 and 9 which we cannot get without concatenation? Any sum, using + and – in any combination, of these 5 digits must be odd. So we have 67 as our other concatenated number, and still need to make an odd number, namely –1, out of 1, 2, 5, 8 and 9. This is potentially possible with 3 odd numbers and the answer:

$1 + 2 + 34 – 5 + 67 – 8 + 9 = 100$ comes out pretty quickly.

We will do many more of this sort of sum in this book so make sure you understand the observations we've made here to significantly cut down the amount of brute force number-crunching you need to do.

035

The first thing to notice here is that the yellow area and the grey area are exactly the same. So if the yellow section has an area of 35, then the entire diagram (yellow and grey) has an area of 70. You can see that this entire diagram can be thought of as 3 whole circles and 2 'middle chunks' of circles with the overlaps removed. The area of a middle chunk is the area of a circle, minus 10. So, 70 = 5 circles minus 20. So the area of 5 circles = 90; that is, a circle has an area of 18.

036

Answer 1) Over 4 days, 36 people would eat 4 × 36 = 144 'people days' of food. An extra 12 people means we now have 48 campers. 144 'people days' of food, for 48 people will last $144/48$ = 3 days.

Answer 2) If 9 women take 36 days to build a house, 1 woman would take 9 × 36 = 324 days. So 12 women would take $324/12$ = 27 days.

Answer 3) If you wanted the house built in 81 days you would need $324/81$ = 4 women.

037

The information 'No two neighbours both had names with an odd number of letters' means the people sit odd (Aaron) - even - odd - even - odd (Eamon) - even - odd - even - then back to odd (Aaron) around the table. You might want to write odd and even under the table spots to help you remember this.

You can write BF beside the two chairs that are the remaining odd spots and CDGH next to the 4 even-lettered chairs.

The information 'Caroline was the only person whose name had the same third letter as the person to their immediate left' means you can cross off C and G from the CDGH immediately to Aaron's left as well as the C from the CDGH immediately to Eamon's right.

Looks like the only seats left for Caroline are on the left side of the table. But what about that seat on the left labelled BF? We now know that person is forced to sit next to Caroline, which mean it can't be Fantine since they share the same last letter. So stick Barry in this seat and Fantine in the other BF seat on the right.

The rule that neighbours can't share last letters means the empty seats on the left are labelled CD (they can't be G or H next to B) and the empty seats on the right are labelled GH (they can't be C or D next to F). Now remember, Caroline is the only person whose left neighbour has the same third letter. So she can't go on Barry's left. By the same token, Gary can't go on Aaron's left.

Now you can place all the even-lettered people and you should get the solution reading the people clockwise from the top: Aaron, Hayley, Fantine, Gary, Eamon, Caroline, Barry, Danielle.

038

For the tables seating 4, we have that the number of girls is a multiple of 4 and the number of boys is a multiple of 4 plus an extra 2. So the numbers of boys and girls could be (2, 36) or (6, 32), (10, 28), (14, 24), (18, 20), (22, 16), (26, 12), (30, 8) or (34, 4).

The result seating 6 students per table gives us a list (6, 32), (12, 26), (18, 20), (24, 14) or (30, 8).

The only overlapping pairs satisfying both conditions are (6, 32), (18, 20), and (30, 8).

So if there are slightly more girls than boys, there are 18 boys and 20 girls in the group.

039

You have to pick the correct number of coins from the total collection, take them to one side and then flip some, or all, of them over. How many coins do you take aside and how many of them do you flip to make sure each group now has the same number of coins showing 'live'?

Choose any 39 of the coins on the table. Take them to one side and flip them all over. The two groups of coins now have exactly the same number that say 'live'.

How does this work? Imagine the 39 coins you selected contained none of the 'live' coins. So your group has 39 'die' and the original group still has exactly 39 'live'. Flip all of yours over and they now both have 39 'live' coins.

Say instead your group of 39 had contained 6 of the 'live' coins from the original group. You have 6 'live' and 33 'die'. The original group now has only 33 'live' coins left. Flip all of your coins over and you'll now have 33 'live' and 6 'die' again, exactly the same as the larger group now has.

In fact, more generally, if you choose n 'live' coins in your random choice of 39, when you flip them all over you will have $39 - n$ 'live' coins in both piles.

So no matter how many or how few of the 'live' coins you choose from the original group, once you flip all of your 39 coins you will have exactly the same number of coins showing 'live' in each group.

040

041

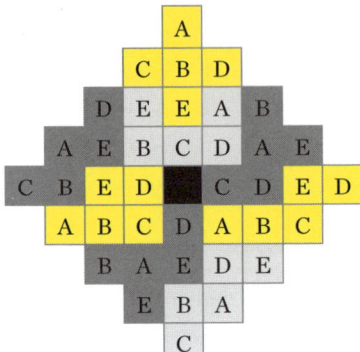

This is the solution I had in mind when I wrote the problem, but doubtless there are more. If you find one, hit me up and let me know. (You can email book@adamspencer.com.au or take a photo, post it and tag me on Twitter (@adambspencer) or Insta (adam_spencer1))

041b

6	2	8	7	9	7	0
7	4	3	5	5	7	7
5	3	3	5	10	10	10
4	9	12	8	12	10	11
1	6	9	12	9	9	11
9	6	5	11	10	11	9
9	6	6	11	9	10	11

15	15	5	6	5	5
5	6	6	4	6	4
4	5	5	5	6	6
6	6	5	5	5	5
5	6	6	4	4	4
5	4	6	5	6	5

042

1	1	3	4	2	2	4	4
2	1	4	1	1	3	4	3
1	4	4	2	4	2	4	2
4	1	3	3	1	1	3	3
4	3	1	3	1	3	2	3
4	3	4	2	2	2	4	1
3	1	3	1	4	4	3	3
4	2	3	3	1	3	2	4

1	2	4	4	1	2	1	2
4	4	1	4	3	3	2	2
2	2	1	1	4	4	2	1
2	4	3	2	4	4	3	2
4	4	2	3	1	3	2	4
1	4	3	3	2	3	2	4
3	3	1	4	1	4	1	2
2	3	1	4	3	4	2	4

043

$5 \times 4 + 5 \times 4 + 3 = 43$

$4 \times 3 \times 2 \times 2 - 5 = 43$

$7 \times 3 \times 2 + 4 \div 4 = 43$

$5 \times 5 \times 2 - 6 - 1 = 43$

$7 \times 6 - 8 + 3 \times 3 = 43$

044

Between them they are laying 66 bricks in an hour, which equates to 11 bricks in 10 minutes or 88 bricks in 80 minutes. So an 88-brick wall would take 1 hour 20 minutes.

With Luigi's interference they now lay a total of 55 bricks in an hour, which equates to 11 bricks in 12 minutes. So 88 bricks would take 8 × 12 = 96 minutes or 1 hour 36 minutes.

044b

Take any grid with three Xs and three Os with the Os in a winning line. There are 8 possible winning lines for O (3 rows, 3 columns and 2 diagonals). For any winning line, the three Os could be placed in 3! = 3 × 2 × 1 different orders. Then there are 6 × 5 × 4 different orders in which the 3 Xs could have been placed. This suggests 8 × 3! × 6 × 5 × 4 = 5760 such patterns.

But we need to exclude any games where X already won with their third move and O then completed their winning line. Think about such a grid. It has to contain a winning line of Xs and one of Os. So neither the Xs or Os can run down a diagonal or there would be no room for the other winning line. They must both be columns or rows. For any of the 6 winning rows or columns of Os which can be played 3! different ways, there are 2 possible winning lines for X which could each be played 3! ways.

So our answer is 5760 − 6 × 3! × 2 × 3! = 5760 − 432 = 5328 possible 6-move games!

045

Above are my solutions to the grids but there are other solutions too; sometimes two solutions involve flipping two pieces within a grid. Let me know how you went!

045b

This can be a bit confusing, so I'll write 5* to mean a 5 in the original grid, which we know could represent any number other than 5. And when I'm writing the actual value I'll just write 5.

The first two rows (from the top) add up to 23 + 14 = 37 and they are missing 4*. So 4* must really be 8. Similarly, rows 1 and 4 add to 23 + 21 = 44 and are missing 7*. So the actual value of 7* is 1. Columns 1 and 2 (from the left) give 1* = 6. Now column 3 gives us 8* = 9, and from row 4, we have 2* + 5* = 7, which combined with column 4 gives us 3* = 7. We're pretty close now. Column 2 now gives 6* = 3 and row 1 gives 9* = 4.

All that is left is 2* and 5*. We know they add up to 7, but we know they are both wrongly labelled, so 2* = 5 and 5* = 2.

So the grid should have been:

4	3	9	7	23
2	1	6	5	14
9	7	6	7	29
5	8	6	2	21
20	19	27	21	

046

This is a tough question so well done if you cracked it. There is enough information in the clues to get an answer, but only just enough. Every clue is vital. Here we go!

- Georgie Grey lives between two houses that begin with a 4. So Grey is 44 or 46.
- [The Wilma White house lives in a corner house] but Gareth Green house does not. So Green is 44, 46, or 48.
- No next door neighbours have surnames that begin with the same letter. So Green is 48 and Grey is 44.

The remaining colours are in 42, 46, 50.

- Brown's Street number is greater than Black's and White's. So Brown is 50.
- The Wilma White house lives in a corner house [but Georgie Green house does not]. So White is 42.

Thus, Black is 46. The residents are Wilma White (42), Georgie Grey (44), Brenda Black (46), Gareth Green (48) and Belinda Brown (50).

047

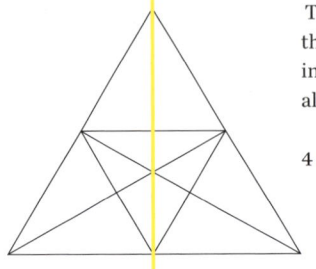

There are 27 triangles in the original diagram. You can also see that once you include a centre line, there are 20 triangles that involve that line. Subtract these 20 triangles from the 47 overall and you again get 27 triangles in the original diagram.

Again, sincere apologies for the pain this caused some of you 4 years ago!

048

Every time we split a side and build a new extension, that side generates 4 new sides in the next snowflake. So there are 3, $3 \times 4 = 12$, $3 \times 4 \times 4 = 48$, $3 \times 4 \times 4 \times 4 = 192$, ... and more generally $3 \times 4^{(n-1)}$ sides for the nth Snowflake.

The Koch Snowflake has infinite perimeter but finite area.

You can see that after the first iteration, each side of the triangle has its length extended by $1/3$ of its original size. So the perimeter has increased by $1/3$ of the original perimeter — or you could say, the perimeter after the first iteration is $4/3$ times the perimeter before the iteration. But you should also be able to see that this is true for every iteration — each time we adjust the snowflake, the perimeter is increased by a factor of $4/3$. So after 3 iterations, the perimeter is $(4/3)^3 = 64/27$ times as big as the original shape. After 50 stages the perimeter is $(4/3)^{50} = 1765780.96326...$ As this process goes on further and further, you can see that the perimeter increases exponentially up to infinity.

Calculating the area change per iteration is much trickier, since you need to know not only the area of each little triangle that gets added, but also how many such triangles are added. Needless to say, you would have to be an utter maths god to calculate that the area after the n iterations is $8/5 - (3/5) \times (4/9)^n$ times as big. But now that I've given you that bit of information, can you tell me what you expect to happen as n approaches infinity? As n gets bigger and bigger, $(4/9)^n$ gets closer and closer to 0 (for example, $(4/9)^{10} = 0.0003...$), which means that the number $8/5 - (3/5) \times (4/9)^n$ gets closer and closer to $8/5 - (3/5) \times 0 = 8/5$. We can therefore say that the Koch Snowflake, which is what we get after infinite iterations, has an area that is $8/5$ times the original area. Pretty cool! (Because it's a snowflake ... geddit?)

048b

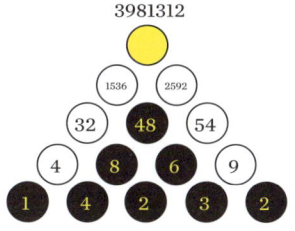

 In this grid, the two numbers below a circle multiply together to give the number above. For example, 8 × 6 = 48. A massive 3,981,312 goes in the top circle. Now 3,981,312 might look like just any old number, but if you notice that the bottom line of the pyramid was all 1s, 2s, 3s and 4s (which are just 22s) it might not surprise you to know that $3,981,312 = 2^{14} \times 3^5$.

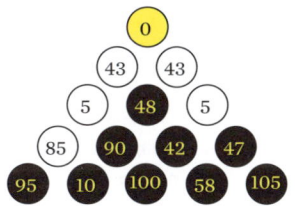

In this grid the number above is given by the difference between the two numbers below. So 90 and 42 give 90 – 42 = 48 but 10 and 100 give 100 – 10 = 90. Ouch! So 0 goes in the top circle.

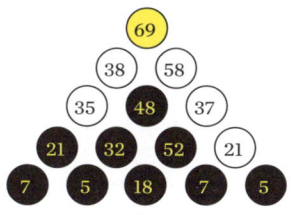

This was a bit of a brute. Add the two numbers together, but REVERSE your answer to get the number above. So 32 + 52 = 84 and 84 reversed is 48.

The completed triangle looks like this, with 69 in the top circle.

049

You can simply smash away with trial and error, trying combinations of ages that add up to 49 until you stumble upon 28 and 21 and realise that Ellie is now 28 and when she was 21 Olivia would have been 14. Ellie's current age 28 is double that of Olivia's age back then, 14. So the answer is Olivia turns 21 today!

A more formal way to get this is to write out some equations.

I'll use E for Ellie, but L for Liv because O for Olivia might look like a zero and get confusing.

We know that E + L = 49.

And if E – L = d (the difference in their ages) the horror sentence 'Ellie is now twice as old as Olivia was when Ellie was the age that Olivia is now' isn't that bad. Ellie is now E and Olivia is now L. So d years ago Olivia was L – d. This gives us E = 2 × (L – d). Replacing that d with E – L tells us E = 2 × (L – (E – L)) = 2 × (L – E + L) = 4L – 2E. So E = 4L – 2E which rearranges to 3E = 4L. Now remember, E + L = 49, so if we multiply both sides by 3 we can see 3E + 3L = 147. Since we just found that 3E = 4L, this is the same as saying 4L + 3L = 7L = 147. So L = 21, and E = 28. Or in other words, this gives us $L/3 = 7$ or $L = 21$. Olivia is 21 and Ellie is 28.

This might make it seem like guesswork is simpler than using equations, but the benefit of equations is they will always give you an answer even when it's too ugly to get by guesswork.

050

$16^3 + 50^3 + 33^3 = 165033$; $166^3 + 500^3 + 333^3 = 166500333$ and $1666^3 + 5000^3 + 3333^3 = 166650003333$. Wow!

050b

What ways can exactly 50 coins give you a dollar?

There are only two combinations of 50 American coins that will add up to $1.00. These are: 40 Pennies, 8 Nickels and 2 Dimes, or 45 Pennies, 2 Nickels, 2 Dimes and 1 Quarter.

In the first case there is a $^{40}/_{50} = {}^{80}/_{100} = 80\%$ chance you drop a penny. in the second case it is $^{45}/_{50} = {}^{90}/_{100} = 90\%$. But you are equally likely to have either combination of coins in your purse. So the odds of dropping a penny is the average of the two, or 85%.

050c

Who'd have thought it'd be a bad idea to make a deal with the devil? Once you agree, ol' Lucifer says, 'You will either give me one of those 50 dollar notes, or you will give me one million dollars.'

Now, to put it mildly, you're stuffed. Look at the logic of the sentence. If you were to give the devil one of the 50 dollar notes, the statement would be true. But here's the catch. In doing so, you would violate the terms of the devil's agreement, that 'if the statement is true, then you keep them both'. So instead, you must give the devil one million dollars.

But what if you gave the devil a different amount? Let's say $60, for instance. This would make the statement false, allowing you to give him back one of the $50 notes without breaking the agreement. But remember the devil's original terms stated that 'if the statement is false, then you give me back just one of the 50 dollar bills'.

The moral here is always pay attention to the terms and conditions. Although you appear to have two choices, the devil's rules mean that, in practice, you have only one: you can't give him a $50 note without breaching the first clause, and you can't give him any amount other than $50 or $1,000,000 without violating the second. You have only one option: pay the devil one million dollars. So, you've learned an expensive lesson.

051

Some of you prime-time fans might have realised immediately that Agent B is wrong: 51 is actually NOT a prime number. It's one of the earliest numbers which people mistakenly call a prime because although 3 × 17 = 51 ... no one remembers their 17 times tables, including Agent B!

Remember that to check whether a number is divisible by 3, all you have to do is add the digits together and check the sum you get is itself divisible by 3. In this case, 5 + 1 = 6, so 51 passes the test.

Agent B thought 51 was prime so he thinks its only factors are 1 and 51, thus the dimensions of his base must have been 1 × 51. Therefore the perimeter of Agent B's base is 2 × (51 + 1) = 104 kilometres. Since Agent A's perimeter was different, his dimensions must be 3 × 17 instead. So, his perimeter is 2 × (17 + 3) = 40 kilometres. The truth is out there!

052

The only order for the numbers that satisfies sentences two and three is, from left to right, 5-3-9. The suits have to read Spade - Spade - Diamond. So from left to right our cards are 5 of spades, 3 of spades, 9 of diamonds. The spades add up to 8.

There are 22,100 ways to choose three cards! How do we get this? Well, there are 52 choices for the first card, then 51 for the second (having reduced the deck by one card), and 50 for the third. That gives us 52 × 51 × 50 possibilities, however if you think about it, you'll soon realise that we've now 'overcounted' by 3 factorial (3!) as that's the number of ways these three selected cards could be arranged. This is because the selection 'Ace of hearts, 5 of clubs, 7 of diamonds' is the same selection as '5 of clubs, 7 of diamonds, Ace of hearts,' and these 3 cards can be selected 6 different ways (namely A75, A57, 7A5, 75A, 5A7, 57A). So we need to divide 52 × 51 × 50 by 3! (that's 6, for those of you playing along at home) to get the answer of 22,100. Cool!

053

The manager said 10 of the first 14 fuses were busted. This means number 14 must be broken or he would have made the more dramatic statement 'Ten of the first 13 are busted'. Ellie chose fuse 14. Sure, if the manager had all this information already, he probably could have changed the fuse himself ... but then he would have done himself out of a place in my puzzle book!

054

A good way to make sure you get all the divisors of a number is to break it down into prime factors. 54 = 2 × 27 = 2 × 3 × 9 = 2 × 3 × 3 × 3. This makes it easier to make sure you pick up all the divisors of 54. They are 1, 2, 3, 6 (2 × 3), 9 (3 × 3), 18 (2 × 3 × 3), 27 (3 × 3 × 3) and 54. So we get the rows:

54
1 2 3 6 9 18 27 54
1 2 2 4 3 6 4 8

And lo and behold, $(1 + 2 + 2 + 4 + 3 + 6 + 4 + 8)^2 = 30^2 = 900$ and $1^3 + 2^3 + 2^3 + 4^3 + 3^3 + 6^3 + 4^3 + 8^3 = 900$. Word up, Liouville!

055

By trial and error you will eventually come up with Dad being 55 and the girls both turning 11 today.

But again let's do it with equations. Let's say Dad is turning D years old today and his gorgeous girls are each turning G. Their ages combine to give 77, so $D + 2G = 77$.

And the girls are G years old, so they were born G years ago. On that day Dad must have turned $D - G$ and we are told this is twice the age that his girls are now combining to be, that is 2G.

So $D - G = 2 \times 2G$, which can be written as $D - G = 4G$ or $D = 5G$.

Combining these two equations we get $D + 2G = 77$ becoming $5G + 2G = 77$ or $7G = 77$. This gives us $G = 11$ so the twins are turning 11 and Dad must be 55.

056

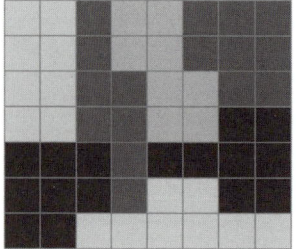

057

The information 'Neither Danielle nor Hayley sat next to men' means they must have sat in between the other two women Caroline and Fantine; so the chain CDHF occurs in some order with the D and H on the inside.

Then using the information 'From Aaron's perspective, Danielle was on the left half of the table' we get the seat that Danielle must be in. If Aaron (A) is in seat 1 and we count clockwise, then D must be in seat 4. So seats 3 and 6 can be labelled CF as these are the only two people who could sit in them. The remaining 3 seats can be labelled BEG.

The information 'No two couples (Aaron/Caroline, Barry/Fantine, Danielle/Eamon, Gary/Hayley) sat the same distance apart' means one couple must be next to each other; one couple must have one seat between them; one couple must be separated by two seats; and, importantly, one couple must sit opposite each other. Well, A is opposite H so neither of them can be involved in the opposite pairing. So either BF or DE are opposites.

What about the couple that's next to each other? It can't involve D or H, since they are stuck between women, and it also can't involve A, since he is stuck between men. This leaves only BF to be the couple with 0 seats between them, and therefore DE is the couple that are opposite each other. So go ahead and label seat 8 with an E. The information 'Danielle sat closer to Barry than she sat to Gary' now tells us that B must be in seat 2 and G in seat 7. Remembering BF is the neighbouring couple, this places F in seat 3 and C in seat 6.

So the answer is, reading clockwise from the top: Aaron, Barry, Fantine, Danielle, Hayley, Caroline, Gary, Eamon.

058

059

$8 \times 6 + 7 + 4 + 2 + 1 - 3 = 59$

$8 \times 6 \times 2 \div 3 + 7 \times 4 - 1 = 59$

$8 \times 7 - 6 \times 2 + 3 \times (4 + 1) = 59$

$(8 + 4) \times 7 - (3 + 2) \times (6 - 1) = 59$

$(7 \times 6 \times 4 + 8 + 2 - 1) \div 3 = 59$

060

061

The Pythagorean primes up to 100 and the squares that sum to them are 5 = 1 + 4, 13 (4, 9), 17 (1, 16), 29 (4, 25), 37 (1, 36), 41 (16, 25), 53 (4, 49), 61 (25, 36), 73 (9, 64), 89 (25, 64) and 97 (16, 81).

062

Grid 1

5	×	2	+	3
−	2	−	8	×
9	×	6	−	3
+	1	×	2	=
5	+	2	=	62

Grid 2

5	×	2	+	3
−	2	−	8	×
9	×	6	−	3
+	1	×	2	=
5	+	2	=	62

Grid 3

5	×	2	+	3
−	2	−	8	×
9	×	6	−	3
+	1	×	2	=
5	+	2	=	62

Grid 4

6	−	4	−	8
×	8	×	5	−
3	×	2	−	5
−	2	×	1	=
9	×	8	=	62

Grid 5

6	−	4	−	8
×	8	×	5	−
3	×	2	−	5
−	2	×	1	=
9	×	8	=	62

Grid 6

6	−	4	−	8
×	8	×	5	−
3	×	2	−	5
−	2	×	1	=
9	×	8	=	62

Grid 7

2	+	4	×	7
+	5	−	1	−
4	×	5	−	9
×	3	+	8	=
3	×	7	=	62

Grid 8

2	+	4	×	7
+	5	−	1	−
4	×	5	−	9
×	3	+	8	=
3	×	7	=	62

Grid 9

2	+	4	×	7
+	5	−	1	−
4	×	5	−	9
×	3	+	8	=
3	×	7	=	62

063

This sort of question is best answered with a diagram and the reason I called the question 'after school ad-venn-tures' was we need our old friend the Venn Diagram.

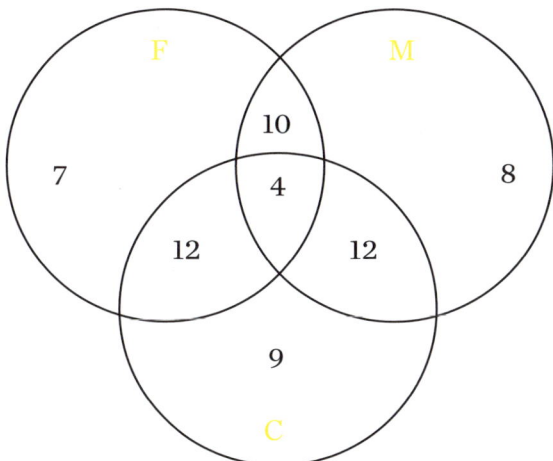

The circles indicate the students who've enrolled in the given activity. If sections of circles overlap, the numbers in the common area are the number of students doing both (or all three) activities.

The secret is to start from the inside and work out. Place the 4 in the centre. The intersection of M and F should add to 14, but there is already a 4 in the centre — so put the remaining 10 in the section above the centre, representing the 4 students doing all 3 activities. Similarly, you can fill out the other intersections with 12 each (16 – 4). Now we have 33 footballers all told and already 12 + 4 + 10 = 26 students doing football and other things. So there are still 7 students to include who are playing football after school but nothing else. Similarly, 8 only do music and 9 only do chess.

Adding all the segments gives 4 + 10 + 12 + 12 + 7 + 8 + 9 = 62 students so there is one child doing none of the after school activities.

063b

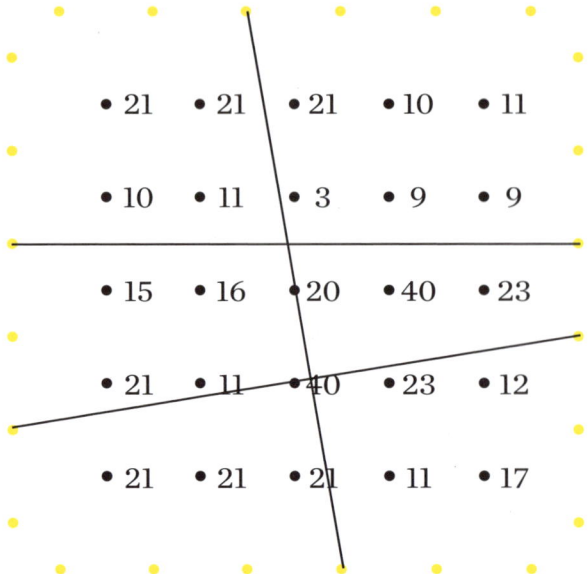

21   21   21   10   11

10   11   3   9   9

15   16   20   40   23

21   11   40   23   12

21   21   21   11   17

064

Corner cubes will have 3 yellow faces and you should be able to see that there are 8 such corner cubes created when I cut the big cube. Edge cubes will have 2 yellow faces and there will be 24 such cubes created. The four centre cubes on each of the big cubes outer faces will have one yellow face and we will have created 24 centre cubes. Don't forget that I will also make 8 'internal cubes' that will have no paint on them at all.

So you are equally likely to draw out a cube with 1 or 2 yellow faces and the odds of doing this for either type of cube is $^{24}/_{64}$ = $^3/_8$ or 37.5%.

065

Obviously 64 can't equal 65, so there must be something dodgy going on.

Look at the slope of the longer side of one of the triangle. It travels 8 units along and goes up 3. So it has a slope of $^3/_8$ = 0.375. The sloped line of the trapezia travels along 5 and up 2, so it has a slope of 0.4. So what looks like a straight line diagonal of the rectangle actually is not a straight line. It kinks slightly. Similarly, comparing the triangle and trapezium on the top half of the rectangle gives another kink. There would actually be a very small gap between these two kinked lines. So the 5 × 13 rectangle isn't actually entirely filled by the 4 yellow shapes. The total area of that gap is the missing unit square.

065b

7	1	7	3	7	6	6	5
1	3	2	6	7	6	1	2
5	2	1	6	1	7	6	5
6	6	5	1	1	7	3	5
4	4	4	2	5	7	7	1
3	1	6	6	6	3	5	3
3	7	1	7	5	2	1	1
1	7	6	4	1	3	6	1

6	7	3	1	6	6	2	2
1	7	6	4	7	1	2	3
6	5	4	1	1	5	3	3
1	4	4	3	6	6	7	4
7	2	5	1	4	5	7	6
6	2	2	6	3	3	7	6
1	1	3	4	5	1	3	5
5	5	4	7	1	4	5	7

If you struggled with these don't fret. I consider these particular puzzles to inhabit 'virtually unsolvable' territory! You can blame my good friend and fellow maths nerd Sean Gardiner — he loves this sort of stuff.

066

066b

2	1	1	2	2	3
3	1	1	1	1	3
3	1	1	1	1	3
2	3	3	2	2	1
2	3	3	2	2	1
2	1	1	2	2	1

2	2	2	3	3	2
3	1	1	1	1	3
3	1	1	1	1	3
2	2	2	1	1	2
2	2	2	1	1	2
1	1	1	3	3	3

1	2	2	2	2	2
3	1	1	3	3	1
3	1	1	3	3	1
2	1	1	2	2	3
2	1	1	2	2	3
2	1	1	1	1	3

067

Think about adding all the numbers in the list –32, –31, –30, ... –2, –1, 0, 1, 2, ... As we hit the positives they start cancelling out their negative partners so by the time we get to –32, –31, –30, ... 0 ... 30, 31, 32 we have a total sum of zero. The next two numbers on the list are 33 and 34 giving us our total of 67. So how many numbers on the list –32, –31, –30, ... 0 ... 31, 32, 33, 34? There are 32 negative numbers, zero and 34 positives making 67 numbers in total.!

067b

The 5 equations involving 67 are $1 + 2 + 34 – 5 + 67 – 8 + 9 = 100$, $12 + 3 – 4 + 5 + 67 + 8 + 9 = 100$, $123 + 4 – 5 + 67 – 89 = 100$, $123 + 45 – 67 + 8 – 9 = 100$ and $123 – 45 – 67 + 89 = 100$.

068

68	51	63	68	47	63
63	47	47	51	51	68
63	68	68	51	51	47
47	51	63	47	68	63
47	63	51	47	68	51
51	68	68	63	63	47

The pattern? Simply rotate the square like so:

68	51
63	47

...

63	68
47	51

...

47	63
51	68

...

51	47
68	63

...

68	51
63	47

... and so on ...

... following the thick black line. I've included the last 2 × 2 grid, so you can see where the one remaining 68 should go.

068b

$66 = 61 + 5 = 59 + 7 = 53 + 13 = 47 + 19 = 43 + 23 = 37 + 29$

069

James is 18, as is Jamal. The full list of information is:

- Jelani does Jump-rope, eats Jaffas, and is 15 (or she could be any age other than 16–19, actually).
- Jitka does Jiu Jitsu, eats Jam doughnuts, and is 16.
- Juanita does Jazz dancing, eats Jelly, and is 17.
- Jamal Jogs, eats Jubes, and is 18.
- Jozsa does Jefferson squats, eats Jawbreakers, and is 19.

070

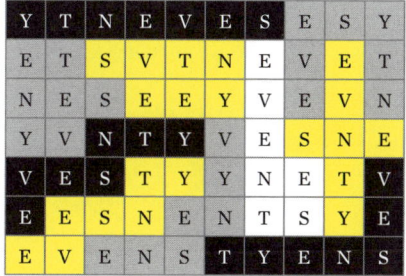

071

$9! = 9 \times 8 \times 7 \times 6 \times 5 \times 4 \times 3 \times 2 \times 1 = 362{,}880$, so $9! + 1 = 362{,}881$ and $362{,}881 = 71 \times 5{,}111$. Now 71 is one more than 70 and 70 is not a multiple of 9, so yes, 71 is a Pillai prime.

For the Brocard problem, $71^2 = 5{,}041$ and $5041 = 5040 + 1 = 7! + 1$, so $(7, 71)$ are Brown numbers. By trial and error you should find the only other two pairs we have found so far, $(4, 5)$ and $(5, 11)$. We've pushed the value of n up to 1,000,000,000 and not found any others, but an absolute proof still eludes us.

071b

These questions get complicated because some answers that may seem acceptable, for example, 4-3 occurring as (X X X X O - O X X O X) can't happen because that shoot-out would be over after Bs fourth shot misses and the score would have been 4-2. Similarly, if the shots went (X X X O X - O X X O X) as soon as team A scores their fourth goal with their final shot, they lead 4-2 and team B can't level up with just one remaining shot. So we see that no matter when Team A misses their goal, Team B has to be on three goals after 4 shots or the shoot-out ends with a 4-2 scoreline. Team A has 5 ways of placing their missing O in a string of 5 shots. Team B must miss their last goal so we are really just placing their other O in one of the first 4 positions. So there are 5 × 4 = 20 different ways the shoot-out can end 4-3.

For a shoot-out to end 4-2, let's think about team A's 4 goals. If they score (X X X X) for their first 4 shots, then team B must miss on their 4th shot for the game to end 4-2. This is because if team B scores on their 4th shot they must have only been on 1 at that stage and when A got their 4th-goal the shoot-out would have finished 4-1. If team B misses their 4th shot we are just placing the other miss in any of the first 3 positions. This can happen 3 ways, O X X or X O X or X X O. So there are 3 ways for the shoot-out to go if A scores their 4 goals in a row.

If A misses one of their first 4 shots and scores with the 5th shot, it is this shot that ends the shoot-out. Team B is locked on 2 goals after 4 shots and can't catch team A so the score is 4-2. This can happen for any combination of team B missing two of their first 4 shots. You can crunch this out by trial and error and find 6 ways it can happen, or observe there are 4 places you can place the first O and 3 places to put the other one. This may suggest 4 × 3 such combinations of shots but notice that we have to halve this number because placing the Os in spots 1 and 3 is the same as placing them in spots 3 and 1, as we have done in earlier counting problems at 004 and 052. There are 4 ways A can miss one of their first 4 and 6 ways B can miss two of their first 4, so 4 × 6 = 24 possible such shoot-outs.

There are 3 + 24 = 27 ways a shoot-out can end 4-2. Phew.

Now take a deep breath as we consider the case of 3-1. A 3-1 shoot-out cannot finish after only 3 shots each because a 3-3 draw is still possible. So let's think about shoot-outs that finish with team A having shot 4 or 5 times.

If A shoots 4 times, scoring 3 times, and B scores once in their first 3 shots but misses their 4th shot the final score is 3-1. A can do this 4 ways and B can do this 3 ways. So a shoot-out like this can proceed 4 × 3 = 12 ways.

If A shoots 5 times, they must have scored with their 5th shot, making the score 3-1 and B doesn't get a 5th shot because the shoot-out is over. Note that if A missed with their 5th shot, then it was already 3-1 when they took this 5th shot and the shoot-out would already have been over. So A is placing two Os into the first 4 positions of (_ _ _ _ X). A has 4 choices for the first O and 3 for the second, but we halve this list of options again because missing 1 and 3 is the same as missing 3 and 1. So there are 6 ways A can miss two of their first 4 shots and score with the 5th. After 4 shots A has scored only 2 goals, so B has to place their one goal in any of the first 4 shots to have the score 2-1 as A prepares to shoot for the fifth time. B can place this X in any of 4 spots. So a 3-1 shoot-out where A scores with their 5th shot can happen 6 × 4 = 24 ways.

So there are 12 + 24 = 36 possible 3-1 shoot-outs that finish within the regulation 5 penalties a side.

072

This is similar to the handshake problem way back at number 6 where we deduced that if 6 people all shake hands with each other there are 5 + 4 + 3 + 2 + 1 = 15 handshakes in total.

But here, when 2 people meet, 2 cards are swapped. So we need to halve 72 and get 36 'connections' taking place. We can then work out how many terms in the equation 1 + 2 + 3 + 4 + ... = 36. You'll see that 1 + 2 + .. + 7 + 8 = 36. So there were 9 strangers at dinner.

Here's another way to think about it:

If there had been 10 people at the dinner, each person would have been given 9 cards, making a total of 10 × 9 = 90 cards handed out. So you can see that at any such meeting, the number of cards handed out will have to equal the product of two consecutive numbers. The way to write 72 like this is 9 × 8, so there must have been 9 people in this case.

073

The middle window is unlocked. If the LEFT window is unlocked then both the LEFT and MIDDLE hints are true and we can't have that. If the RIGHT window is unlocked, then both the MIDDLE and RIGHT hints are true, again not possible. But if the MIDDLE window is unlocked then LEFT and MIDDLE are lies and the RIGHT hint is the correct one.

You might have noticed that the left and right windows' messages actually contradict each other ... so one of them has to be true, and the other has to be false. Either way, the middle window also has to be false ... so we immediately know that middle window is the unlocked one!

073b

The gold is in the YELLOW suitcase.

If the gold is in the YELLOW suitcase then the YELLOW and GREY clues are true and the WHITE clue is false. If the gold is in the GREY case, all the clues are false and if the gold is in the WHITE case all the clues are true. So the YELLOW case is the only place the gold can be to leave at least one clue true and at least one clue false.

074

74	97	97	74	82	74
82	28	28	82	28	97
82	28	28	82	28	97
74	97	97	74	82	74
97	28	74	82	74	97
74	82	97	28	82	28

Reflect the square through its 4 axes of symmetry, like so:

74	97
82	28

...

97	74
28	82

...

82	74
28	97

...

28	97
82	74

...

28	82
97	74

... and so on ...

... following the thick black line.

075

The fraction has to equal 25, so we must want the top to come out to 75 and the bottom to equal 3. The closest we can get to 75 with two digits is 9 × 8=72, so let's make 75 from 9 × 8 + 3. Then we have the digits 1, 2, 4, and 6 left to wrangle into a 3.

After a bit of practice, you should find 3 = 24 ÷ 6 – 1. So the answer is:

$$100 = 75 + \frac{(9 \times 8 + 3)}{(24 \div 6 - 1)}$$

075b

Michelle Mailer handed over two 50 cent pieces as her dollar.

If she'd wanted a 25 cent stamp, she would only have given one 50 cent coin to the woman at the counter.

076

The full list of solutions, with the '76-ers' in bold are:

98 – 76 + 54 + 3 + 21 = 100
9 – 8 + 76 + 54 – 32 + 1 = 100
98 + 7 + 6 – 5 – 4 – 3 + 2 – 1 = 100
98 – 7 – 6 – 5 – 4 + 3 + 21 = 100
9 – 8 + 76 – 5 + 4 + 3 + 21 = 100
98 – 7 + 6 + 5 + 4 – 3 – 2 – 1 = 100
98 + 7 – 6 + 5 – 4 + 3 – 2 – 1 = 100
98 + 7 – 6 + 5 – 4 – 3 + 2 + 1 = 100
98 – 7 + 6 + 5 – 4 + 3 – 2 + 1 = 100
98 – 7 + 6 – 5 + 4 + 3 + 2 – 1 = 100
98 + 7 – 6 – 5 + 4 + 3 – 2 + 1 = 100
98 – 7 – 6 + 5 + 4 + 3 + 2 + 1 = 100
9 + 8 + 76 + 5 + 4 – 3 + 2 – 1 = 100
9 + 8 + 76 + 5 – 4 + 3 + 2 + 1 = 100
9 – 8 + 7 + 65 – 4 + 32 – 1 = 100

–9 + 8 + 76 + 5–4 + 3 + 21 = 100
–9 + 8 + 7 + 65 – 4 + 32 + 1 = 100
–9 – 8 + 76 – 5 + 43 + 2 + 1 = 100

077

The information 'Danielle sat opposite a female' means we must have an opposite of DC or DH.

The information 'Eamon sat closer to Fantine than he sat to Caroline' means neighbours EC are not possible.

The information 'Barry and Hayley sat the same distance from Gary, with Barry closer to Gary's right' leaves a lot of possibilities. They could all sit next to each other, for example, seats 2, 3 and 4 (where Aaron is in seat 1 and we are numbering clockwise) could be BGH in that order, or seats 6, 7 and 8 could also read BGH. But each time, D would have to sit opposite to H and that would leave E next to C which is not allowed.

But BGH could sit in any three of the four even-numbered seats and they'd still satisfy 'Barry and Hayley sat the same distance from Gary, with Barry closer to Gary's right'. No matter which three even-numbered seats they take, H will be opposite B, and so D's opposite will have to be

C, with DC sitting in seats 3 and 7 in some order. But if C is in an odd-numbered seat and H is in an even-numbered seat, it is impossible to satisfy 'There were an odd number of chairs between Caroline and Hayley'.

So BGH must sit with two chairs between them, for example, in seats 4, 7 and 2. This is the only arrangement possible with two seats between them because don't forget, 'Barry, Danielle, and Fantine sat together in some order'.

Then noting that D has to sit opposite C or H and remembering 'Eamon sat closer to Fantine than he sat to Caroline', you soon find the solution, reading clockwise from seat 1: Aaron, Hayley, Eamon, Barry, Fantine, Danielle, Gary, Caroline.

078

The concatenation 78 occurs in the following three equations:

$1 + 23 - 4 + 5 + 6 + 78 - 9 = 100$, $1 + 2 + 3 - 4 + 5 + 6 + 78 + 9 = 100$ and one you might not instantly find because it starts with a negative sign $-1 + 2 - 3 + 4 + 5 + 6 + 78 + 9 = 100$.

078b

Look at the answer to 6 to see the connection between triangular numbers and handshakes; obviously the same applies for fist bumps and high fives.

So we need to find 3 triangular numbers that add up to 78, where your options for triangular numbers are: 1, 3, 6, 10, 15, 21, 28, 36, 45, 55, and 66. A bit of work gives you 3 possibilities: $78 = 66 + 6 + 6 = 36 + 36 + 6 = 36 + 21 + 21$. This corresponds to the number of people being (11, 3, 3), (8, 8, 3), or (8, 6, 6). But only in the middle case can the number of adults be less than both the number of girls and the number of boys. So there must be 8 girls, 8 boys, and 3 adults.

079

$6 \times 4 \times 3 + 5 + 2 = 79$
$7 \times 6 \times 4 \div 2 - 5 = 79$
$5 \times 4 \times 4 - 6 \div 6 = 79$
$9 \times 2 + 8 \times 7 + 5 = 79$
$8 \times 5 \times 4 - 9 \times 9 = 79$

080

Sixteen. The numbers are all factors of 80, apart from the trivial factors of 1 and 80 itself.

080b

The sum of EIGHTY is 74 and 80 is the first number whose letter sum is less than itself.

081

This is a tough question so well done if you got it!

You can work out pretty quickly that the last digits of the numbers involved in the fractions have to be 3 and 7 or 4 and 6.

Trial and error and the hints I've given you get you to: $100 = 81 + {}^{5643}/_{297} = 81 + {}^{7524}/_{396}$.

082

82	−	28	÷	2	=	68
÷			−		+	
2	+	8	+	28	=	38
−			÷		×	
8	×	2	×	2	=	32
=			=		=	
33		24		58		

82	−	2	−	28	=	52
−			×		÷	
8	×	8	÷	2	=	32
−			−		+	
8	×	2	+	82	=	98
=			=		=	
66		14		96		

2	+	28	−	8	=	22
+			×		×	
82	÷	2	+	8	=	49
−			÷		+	
28	×	8	÷	8	=	28
=			=		=	
56		7		72		

For the last grid, it's easiest to start with the middle row, which can only be $8 + {}^{82}/_2$ or ${}^{82}/_2 + 8$. If the first case is true, the third column can only be $82 - 2 - 8$ … but then it is impossible for the top row to make 22 using an 82 in its equation.

So the middle row must be ${}^{82}/_2 + 8$. See how you go from there …

Notice that in the solution for the third grid, the third column could be $8 × 8 + 8$ OR $8 + 8 × 8$, and the third row could be $28 × {}^8/_8$ OR ${}^{28}/_8 × 8$ OR $28 + 8 - 8$ OR $28 - 8 + 8$.

082b

Something that makes the 82 version of this problem a bit easier than the 81 version is that the numerator has to be divisible by 18, so it needs to be even, and divisible by 9. You might remember the divisibility test for 9: a number is divisible by 9 if its digits add up to a multiple of 9.

Along with the hint about consecutive digits, this means the numerator must be made out of the digits 3, 4, 5, 6 in some order (the other possible run is 6, 7, 8, 9, but 8 is not available to us). And because the number must be even, the last digit can only be 4 or 6. See how you go from there … the answer is: $100 = 82 + 3546/197$

083

Epsilon must be between 46 and 66 years old now because he was born strictly between 1951 and 1973, and Delta is at most 45.

So by 2027 Epsilon will be between 55 and 75. Write out a list of the prime numbers up to 79, the first prime after 75. Underneath each of them write that number less 9. Now you need to find 3 consecutive primes that add up to a prime number between 55 and 75. The two options are 17, 19 and 23 which add to 59 and 19, 23 and 29 which add to 71. So the children are either 8, 10 and 14, with a 32-year-old mother and 50-year-old father or 10, 14 and 20 with a 44-year-old mother and 62-year-old dad. If Delta is 44 she should have been born in 1974 only just missing the window 1951–73 but if she is 32 she wasn't born until 1986, well outside of the 1951–73 window. So Delta is 44, Epsilon is 62 and their age difference is 18 years.

On the other hand, if you're a bit of a maths guru, you might have also noticed that lots of the information I gave you was actually irrelevant to answering the question. If we assign the names a letter each as below, we can see that:

A + B + C = D, and A + 9 + B + 9 + C + 9 = E + 9

Therefore ... D + 18 = E

Note that with this solution, we didn't even need to use Epsilon's precious primes!

084

The 30 cm^2 face could be, in centimetres, 1 × 30, 2 × 15, 3 × 10 or 5 × 6. The 70 cm^2 could be 1 × 70, 2 × 35, 5 × 14 or 7 × 10 and the 84 cm^2 side could be 1 × 84, 2 × 42, 3 × 28, 4 × 21, 6 × 14 or 7 × 12. But the faces share common edges so you should be able to see the only combination that matches up all 3 faces is 5 cm × 6 cm × 14 cm. So the volume of the box is 420 cm^3.

If you're into your algebra, you might also have tackled it like this, without 'guess and check':
Let the side lengths be a, b and c. Then:

$ab = 30$

$ac = 70$

$bc = 84$

The first two equations combine to give $a^2 = 2100/bc = 2100/84 = 25$.
So a = 5, b = 6, c = 14.

085

The 3 groups of four that sum to 20 are: 1, 2, 8, 9; 1, 4, 7, 8; and 2, 4, 6, 8. If you try putting 1, 2, 8, 9, 10 into the corners of a pentagon, though, you'll soon see one of the sides will contain 10 and 9, 10 and 8, or 9 and 8 ... so that side will always have a total greater than 17.

If you try putting the other 2 groups into the pentagon's corners, you should eventually find these 2 solutions:

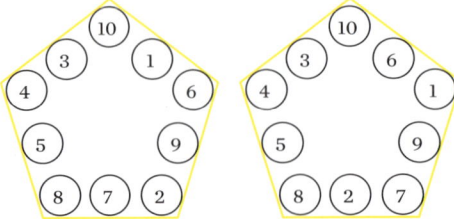

For the case of the sides all adding to 16, the circles read off in order, starting from the top corner, 1, 5, 10, 2, 4, 9, 3, 6, 7, 8 and 1, 10, 5, 2, 9, 4, 3, 6, 7, 8. You can see that these two solutions are just the mirrors of the two solutions for 17 where 10 swaps with 1, 9 with 2, and so on!

For the cases of 15 and 18, I'll show you how to prove the 15 case is impossible, and see if you can then find a very similar argument for showing the 18 case is also impossible.

As with the other cases, this time the corner numbers have to add up to 20 (because 15 × 5 – 55 = 20). Let's say 3 of the corner numbers are between 5 and 10. Then the smallest possible sum our set of 5 numbers could take is 1 + 2 + 5 + 6 + 7 = 21, which is already bigger than 20. This means that at most two of the corner numbers are between 5 and 10. So at least 3 of the corner numbers are between 1 and 4. But if you pick any 3 numbers between 1 and 4, you can see that there will always be a pair whose sum is 5 ... which is bad news for us, because a pair adding to 5 means there's a sub-triple adding to 15, which we saw is not allowed to happen when discussing the 17 case.

So ... case closed! There's no way to pick the 5 corner numbers, so there are no pentagons with sides whose sums are each 15.

086

If Felipe only needs these 3 questions, Nadja must have answered No, Yes, No and lives in number 64.

To see this, it should be obvious that she must live in a perfect square or there would just be too many possible houses to consider that we couldn't eliminate all but one with only two other questions.

So assuming answer number 2 is a Yes. The 3 answers Yes, Yes, Yes leaves us two odd squares that are below 43, namely 9 and 25, that could be Nadja's house. Similarly No, Yes, Yes leaves 4, 16 and 36 as possibilities. Yes, Yes, No leaves two odd squares that are between 43 and 86, namely 49 and 81. But No, Yes, No leaves the only even square within that range — namely 64.

086b

No, the laces do not cost $1. If they did, the shoes would have cost $85 which is only $84 more than the laces. The shoes actually cost $85.50 and the laces just 50c. Bargain!

087

1	1	1	2	2	2
3	4	4	3	3	2
3	4	4	3	3	2
1	2	2	1	1	4
1	2	2	1	1	4
2	3	3	4	4	2

2	3	3	4	4	3
4	1	1	2	2	1
4	1	1	2	2	1
2	4	4	3	3	1
2	4	4	3	3	1
3	1	1	2	2	3

3	4	4	2	2	1
1	1	1	3	3	4
1	1	1	3	3	4
3	4	4	2	2	1
3	4	4	2	2	1
1	2	2	3	3	2

088

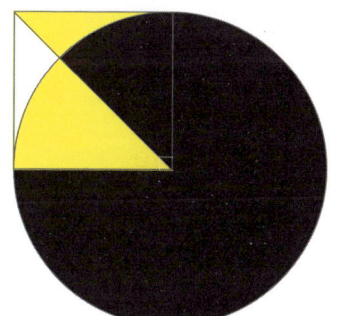

If you reflect the top yellow segment through the diagonal of the square, you can see the yellow triangle has area half that of the square. So the yellow area = $1/2 \times 88 \times 88 = 3872$ square centimetres.

088b

By joining the centres of the 6 circles, we get a way to calculate the region. The yellow outline is a hexagon, and each of its sides is made of two radii — that is, it's a regular hexagon with side-lengths 88 + 88 = 176 cm. You can also see that the 3 light grey components can come together to make a circle, as can the 3 dark grey components, and each circle has radius 88 cm. So the area of the yellow region is whatever the area of the hexagon is, minus the area of the 3 circles (white, light grey, and dark grey).

Here's where the hardcore geometry comes in.

A regular hexagon is made up of 6 regular triangles with the same side-length, so let's find the area of an equilateral triangle whose sides are each length 176 cm. The trigonometry nerds reading this will know that the area can be expressed as $(1/2) \times 176 \times 176 \times \sin(60°)$, and $\sin(60°) = \sqrt{3}/2$. So the hexagon's area is $6 \times (1/2) \times 176 \times 176 \times (\sqrt{3}/2)$.

Now the area of each circle is $\pi \times 88^2$, so bringing all this together, the area of the yellow region is:

$6 \times (1/2) \times 176 \times 176 \times (\sqrt{3}/2) - 3 \times \pi \times 88^2 = 74992.528...$ square centimetres. Nice!

089

For 4-digit numbers there are 5 possible different answers. These are: 0, 990, 9999, 10890, and 10989.

For 5-digit numbers, there are also only 5 possible different answers. These are: 0, 10890, 99099, 109890, and 109989.

How many of these did you find?

089b

$1 + 2 + 34 - 5 + 67 - 8 + 9 = 100$
$12 + 3 - 4 + 5 + 67 + 8 + 9 = 100$
$123 - 4 - 5 - 6 - 7 + 8 - 9 = 100$
$123 + 4 - 5 + 67 - 89 = 100$
$123 + 45 - 67 + 8 - 9 = 100$
$123 - 45 - 67 + 89 = 100$
$12 - 3 - 4 + 5 - 6 + 7 + 89 = 100$
$12 + 3 + 4 + 5 - 6 - 7 + 89 = 100$
$1 + 23 - 4 + 5 + 6 + 78 - 9 = 100$
$1 + 23 - 4 + 56 + 7 + 8 + 9 = 100$
$1 + 2 + 3 - 4 + 5 + 6 + 78 + 9 = 100$
$-1 + 2 - 3 + 4 + 5 + 6 + 78 + 9 = 100$

And for the decimal sum observe the following. If we took 1 and 2 and made it 1.2 we'd need to add it to 7.8 to get a whole number. The 8 is already used in 89 so 1.2 can't be part of our answer. Also 4.5 doesn't have another 0.5 to add to make an integer. The only choices would be 2.3 + 6.7 or 3.4 + 5.6. This doesn't leave much calculating to get our answer $1 + 2.3 - 4 + 5 + 6.7 + 89 = 100$.

089c

I'll use the notation 89 $(-,199)$ to represent $5 \times 89^2 - 4 = 199^2$. For the Fibonacci numbers up to 89 you get 1 $(-,1)$ and $(+,3)$, 2 $(-,4)$, 3 $(+,7)$, 5 $(-, 11)$, 8 $(+,18)$, 13 $(-,29)$; okay, that one was a bit tough and it gets hard from now without using a calculator, kudos if you did. 21 $(+,47)$, 34 $(-,76)$ and 55

(+,123). A trick that might have helped you with the larger numbers is to notice that whether you want to add 4 or subtract 4 alternates with each successive Fibonacci number.

090

Obviously one of the hairdressers, let's say Bob Bob, will have to perform 2 haircuts. This will take 180 minutes. This leaves Henrietta to perform one haircut, a shave, shampoo and blow dry which will take 90 + 30 + 30 + 30 = 180 minutes.

So the quickest time would be 180 minutes if we can find a combination that works. And we can. Salome can't be blown dry until after her cut so Bob should cut her hair first. Henrietta can do the treatment and shave in this time then start on a haircut for one of these people. Sophia needs an hour's wait so let's treat her hair first, then shave Sergio and when that's done begin on Sergio's haircut. Once Bob has finished Salome's haircut he can start cutting Sophia's hair, which has been treated for an hour now. Once Henrietta has finished cutting Sergio's hair she only has to blow dry Salome and we are done and dusted in 180 minutes.

Which is about the amount of time I spend on my hair in total every 4 months!

091

Similarly to the case for 82, here the top number must be a multiple of 9, so we know its digits must add to a multiple of 9. My hint told you 2 and 5 were already in the numerator, so that already gives us a digit sum of 7. It's impossible to throw in two more digits and have the sum equal 9 or 27, so the digit sum must be 18. We need to add two more digits that sum to 11 ... so they can either be 3 and 8, or 4 and 7.

After a fair bit of patience, you should have discovered the 3 solutions: $100 = 91 + {}^{5742}\!/_{638}$, $100 = 91 + {}^{5823}\!/_{647}$, and $100 = 91 + {}^{7524}\!/_{836}$. Phew!

091b

This isn't too hard to work out. In a 5-move game, X has played just 3 moves (a winning 3 in a row) and O has gone on two of the other squares not on the winning line. There are 8 possible winning lines (3 rows, 3 columns and 2 diagonals). Each of those winning lines can be played $3 \times 2 \times 1 = 3! = 6$ different ways. And for each winning line there are 6 unused squares. O must go in two of these and can do that 6×5 different ways.

So the total number of possible games at the end of the 5th move is $8 \times 3! \times 6 \times 5 = 1440$.

092

Whitney played black against Cameron and White against Kyeema at exactly the same time.

Against Kyeema she copied Cameron's moves ... against Cameron, she copied Kyeema's. So Cameron and Kyeema were effectively playing the same game against each other on both boards. Therefore Whitney must win one game and lose one game, or draw both games. Either result is better than Blake Black's two losses.

093

Trick question! The lengths AB, AC and BC are all diagonals of squares of the same size, so AB = AC = BC. This means all 3 angles of the triangle are equal to 60 degrees. And since this argument obviously doesn't use the edge lengths of the cube at all, it doesn't matter how big or small the cube is — the angles remain equal for any new-sized cube.

094

$100 = 94 + {}^{1578}/_{263}$

095

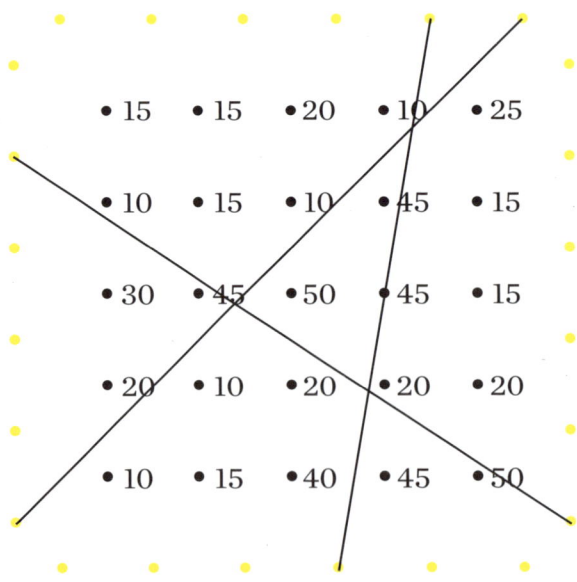

096

If you think back to previous problems like this where we used divisibility tricks, in this case we know the numerator has to be divisible by 4, which is the same as saying its last two digits have to be divisible by 4. Using that as a starting point, see if you manage to reach these 3 answers:

$$100 = 96 + {}^{2148}\!/_{537}$$
$$= 96 + {}^{1428}\!/_{357}$$
$$= 96 + {}^{1752}\!/_{438}$$

096b

5	×	4	—	8
—	2	×	2	×
6	+	8	×	7
—	5	—	3	=
3	—	9	=	96

5	×	4	—	8
—	2	×	2	×
6	+	8	×	7
—	5	—	3	=
3	—	9	=	96

5	×	4	—	8
—	2	×	2	×
6	+	8	×	7
—	5	—	3	=
3	—	9	=	96

1	+	6	—	1
×	9	—	2	+
6	+	8	×	5
—	6	×	4	=
2	—	7	=	96

1	+	6	—	1
×	9	—	2	+
6	+	8	×	5
—	6	×	4	=
2	—	7	=	96

1	+	6	—	1
×	9	—	2	+
6	+	8	×	5
—	6	×	4	=
2	—	7	=	96

3	—	2	+	1
×	6	+	5	×
4	+	9	×	3
+	2	×	1	=
8	—	6	=	96

3	—	2	+	1
×	6	+	5	×
4	+	9	×	3
+	2	×	1	=
8	—	6	=	96

3	—	2	+	1
×	6	+	5	×
4	+	9	×	3
+	2	×	1	=
8	—	6	=	96

097

$6 \times 5 \times 3 + 8 \div 2 + 4 - 1 = 97$

$8 \times 4 \times 3 - 6 \div 2 + 5 - 1 = 97$

$(8 \times 5 - 6 \times 4) \times 3 \times 2 + 1 = 97$

$(6 + 5) \times (8 - 1) + 4 \times (3 + 2) = 97$

$(8 + 4 + 1) \times (5 \times 2 - 3) + 6 = 97$

097

Draw a round table with 8 circles evenly spaced around it.

Place Aaron at the top in seat 1 and number them clockwise. The information 'Each person sat next to both a man and a woman' tells us that the pattern must be either MWWMMWWM, or MMWWMMWW, where Aaron is the first M. You might want to draw two tables here and label the seats in each of these M&W patterns. The information 'From Eamon's perspective, Caroline was on the left side of the table, and Barry was on the right' tells us importantly that Eamon can't be opposite Barry. Coupling this with 'Barry didn't want to sit next to, nor face, Gary' tells us that Barry must be opposite Aaron. We also know that Barry must be next to Eamon, on his right, so this gives us the MMFFMMFF pattern of chairs starting with Aaron and puts Eamon in chair 5 , leaving chair 2 for Gary. We can then use the information 'Each couple (Aaron/Caroline, Barry/Fantine, Danielle/Eamon, Gary/Hayley) sat together' to finalise the arrangement.

Clockwise from the top: Aaron, Gary, Hayley, Fantine, Barry, Eamon, Danielle, Caroline.

Bonus point: the restaurants were at numbers 37, 57, 77 and 97 ... what we number nerds call an 'arithmetic progression.'

098

Because the pirates don't trust each other to make any deals among themselves, Pirate A can give 1 coin to Pirate C, 1 coin to Pirate E, and keep 98 coins for himself. If he does, Pirates C and E will vote with him, since they know they both will end up with nothing if they don't!

As a result, the coins will be distributed: A:98, B:0, C:1, D:0, E:1. But how do pirates C and E know pirate A's offer is the best they'll get?

Working backwards: say there are only two pirates left: D and E. Pirate D will hoard all 100 coins for himself, knowing that his vote will give him the 50% he needs. So, D:100, E:0.

If Pirates C, D and E are divvying up the coins, the distribution would be C:99, D:0, E:1. Why? Pirate C will give 1 coin to Pirate E, and 0 coins to Pirate D. Pirate E will have to vote with Pirate C, since he knows that if they hurl C overboard, D will give him diddly-squat.

Now, with four pirates, B will give 1 coin to D, and D will vote along with him, knowing that if B goes overboard, C will give him nothing. B and D together give the proposal the 50% vote it needs to pass. The reason Pirate B won't give 1 coin to Pirate E instead is because he knows he can't get any more coins — but Pirate E will vote to toss B overboard just for fun if he thinks his coins will be the same from C. So, B:99, C:0, D:1, E:0.

So you can see that, because of all this, A can safely offer C just 1 single, solitary coin, and E the same. If the two pirates vote to make him walk the plank, they'll both be left with nothing when B gets his turn to distribute the coins. Final distribution? A: 98, B:0, C:1, D:0, E:1.

Aaaarh!

099

Yep, this was EXCEPTIONALLY hard. Here's how I cracked it.

Label the dice A B C D E F as they appear down the page, A to the left of B, then C to the left of D and in the last roll E to the left of F.

A and E both contain 69, which is the same when you view it upside down so are the same die. That means B and F must be the same die, and that gives us at least 5 different faces of this die: 16/96, 18/81, 66/99, 68/89, and 89/98. D must contain both 19 and its flip 61. So D can't be the same die as B and F, because it would have to have more than 6 different numbers on its faces (the 5 I just listed, plus 19 and 61). So D must be the same die as A and E.

This would mean A D and E are one die and B C and F the other.

Notice that the 86/98 face on die B cannot be the same as the 86/98 face on die C, because if you try to rotate one die to match the other, the remaining visible faces do not match up. So the die corresponding to dice B, C, and F must contain both 86 and 98 as separate faces. From now on, let's just call the two dice A (which is the same as D and E) and B (which is the same as C and F).

A has faces 19, 61, 69 and 18/81, 68/89 and 66/99.

B has faces 86, 98 and 16/91, 68/89, 66/99 and 18/81.

From the die sums we know the first pair have bases summing to 105. The only way to do this is 19 + 86 or 89 + 16. But the 16/91 on B is visible so it can't be on the base. It must be 19 on the base of A and 86 on the base of B. We both knew these numbers were on these dice already.

The second sum of 149 could be given by 68 + 81 or 81 + 68. But the 68/89 in A is visible so not on the base. The only option is 81 on the base of A and 68 on the base of B. This gives us new information about some of the numbers on the dice.

So A now contains 19, 61, 69, 81, 89 and 66/99

and B contains 18, 68, 86, 98 and 16/91 and 66/99.

Now look at the third roll of the dice. By rotating the 3 die images that make up A you can see that the base is 61.

Similarly, by comparing the 86/98s on the 3 die images that make up B the base of B has to be the 98.

So the sum must be 61 + 98 = 159.

By the way, you could have avoided some of the work above if you considered the 4 numbers I offered you as the only possible sums. For example, as soon as you knew the bottom of die A is 61, you could be sure the sum has to be 159, because the other three sums are rendered impossible (for example, if the sum were 100, die B would have to have 39 on the bottom, which is not on any of its faces).

If you want to go one step further you can see that the 'ambiguous faces' of the dice are a 66/99 on each and the 16/91 on B.

Phew! Fiendish indeed!

100

$1! \times 2! \times 3! \times 4! \times 5! \times 6! \times ... \times 99! \times 100! = (1! \times 1!) \times 2 \times (3! \times 3!) \times 4 \times (5! \times 5!) \times 6 \times ... \times (99! \times 99!) \times 100 = (1! \times 1!) \times (3! \times 3!) \times (5! \times 5!) \times ... \times (99! \times 99!) \times 2 \times 4 \times 6 \times ... \times 100 = (1! \times 3! \times 5! \times ... \times 99!)^2 \times (2 \times 1) \times (2 \times 2) \times (2 \times 3) \times (2 \times 4) \times ... \times (2 \times 50) = (1! \times 3! \times 5! \times ... \times 99!)^2 \times 2 \times 2 \times 2 \times ... \times 2 \times 1 \times 2 \times 3 \times ... \times 50 = (1! \times 3! \times 5! \times ... \times 99!)^2 \times 2^{50} \times 50!$

But 2^{50} is just 2^{25} squared, so drop the 50! and the remaining product is equal to $(1! \times 3! \times 5! \times ... \times 99! \times 2^{25})^2$

Ouch, that hurt. You can go lie down now. Well done.

100b

For every chick in the circle, the chances of being pecked from the right is 0.5, and not getting pecked from the right is 0.5. The same for getting pecked or not pecked from the left. So the odds of not getting pecked at all is $0.5 \times 0.5 = 0.25$ (the same odds as getting 'double pecked' in fact). So across the 100 chicks we would expect 25 to not get pecked.